Henry Lawson

A Manual of Popular Physiology

Being an Attempt to Explain the Science of Life in Untechnical Language

Henry Lawson

A Manual of Popular Physiology
Being an Attempt to Explain the Science of Life in Untechnical Language

ISBN/EAN: 9783744643573

Printed in Europe, USA, Canada, Australia, Japan

Cover: Foto ©berggeist007 / pixelio.de

More available books at **www.hansebooks.com**

A MANUAL

OF

POPULAR PHYSIOLOGY:

BEING

AN ATTEMPT TO EXPLAIN THE SCIENCE OF LIFE
IN UNTECHNICAL LANGUAGE.

BY

HENRY LAWSON, M.D.

NEW YORK:
G. P. PUTNAM'S SONS.
1873.

TO

E. V.

IN REMEMBRANCE OF MUCH KINDNESS,

𝕿𝖍𝖎𝖘 𝖑𝖎𝖙𝖙𝖑𝖊 𝖁𝖔𝖑𝖚𝖒𝖊

IS AFFECTIONATELY DEDICATED

BY

THE AUTHOR

PREFACE.

IN the following chapters it has been attempted to set before the public, in as easy a style as possible, the more interesting results of physiological research. Dry details have been avoided, except where the nature of the subject rendered their introduction inevitable; whilst the grand general principles and laws of the Science have been enunciated in language as devoid of technicality as it seemed advisable to employ. In no case has truth been distorted to attain simplicity; and withal the volume (it is hoped) embodies the latest discoveries in the branch of knowledge upon which it treats.

This work is not addressed to the man of Science; nor is it supposed that students proceeding to degrees in Medicine will devote much attention to it. There is, however, a large class, in which the author is happy to say he once ranked, that, although not professedly scientific, is glad to profit by the gleanings from the great field of Biology. For it, this little volume has been written; and if it inspires even one person with a desire to embrace the cause of Science, the Author will be amply satisfied.

In order that those who wish to pursue the subject as a *spécialité* may do so without difficulty, the names of the more important treatises on Physiology and on other sciences which relate to it have been appended to the volume. Should other references be needed, they may be found in the Medico-Chirurgical and Natural History Reviews, the Archives of Medicine, and the valuable bibliographical compilation of Professor Carus.

There is an old and familiar proverb regarding the strangulation of that member of the canine genus which has through some mischance been deprived of its fair reputation. The Author has no inclination to institute unseemly analogies; nevertheless it appears to him that, to the minds of some English *savans,* popular Science merits the fate of the unlucky being to which he has alluded. He regrets that such a state of things exists, and hopes that ere long those who so zealously decry untechnical Science, on the principle that half a loaf is *not* better than no bread, may be relieved of the painful hallucination under which they now labour. Are we to assume that because a man is unable to read "Faust" in the language of Goethe, he is therefore unwarranted in enjoying a translation of that sublimest of German writings?

In conclusion, the author would remark that in the footsteps of BREWSTER, GOSSE, JONES, HUXLEY, and LANKESTER, who have all written for the people, no man need blush to follow.

QUEEN'S COLLEGE, BIRMINGHAM,
June 10*th*, 1863.

POPULAR PHYSIOLOGY.

CHAPTER I.

INTRODUCTION.

Relation of other Sciences to Physiology—Three kinds of Natural Objects—Distinction between Animals and Plants—Nature of an Element—Action of Vegetables on the Earth and Air—Dependence of Animals upon the Plant-World—Passage of Fluids through Membranes—Osmose—Endosmose—Exosmose.

In order to study with advantage the higher departments of Physiological Science it is almost essential that one should be acquainted with the general principles of Anatomy, Zoology, Botany, Chemistry, and Experimental Physics; but as in the following pages the rudiments of Physiology have alone found a place, it will not be necessary for the reader to possess a knowledge of these comprehensive and difficult branches of learning. However, since it is impossible that any clear description of the processes gone through by the human body, could be given without some reference to other sciences, I shall devote this chapter to a few remarks on collateral information bearing on Physiology.

A moment's reflection will suffice to convince anyone that each object by which he is surrounded is one of three kinds: thus, let the reader suppose himself in a balloon, some thousand feet above the earth's surface, and on looking down, say, in the most general way possible, how many sorts of things he sees; or, let him imagine that he has got to arrange the various objects he beholds in such a manner that there shall be but *three* divisions, and yet that all the contents of each division may be of the same description, and I have no doubt he will observe, that the three groups are minerals, vegetables, and animals. So far so good; having bundled up the different materials in this way, we shall leave the first to the chemists and geologists, whilst our friends the botanists shall share the vegetables with us, and the zoologist shall classify the animals, so that we shall have no difficulty in pitching upon the beasts we may require for our experiments. If it be asked, how are we always to decide whether an object belongs to any particular class; or how shall we maintain our rights, and prevent those speculative ay and combative fellows, the geologists, from trespassing upon our premises? I must reply that here, even in the present day, we are "in a fix," and notwithstanding that loud-sounding essays have been written, and terrific, jaw-breaking Greek names have been coined, we have yet arrived at very little improvement on the statement of Linnæus, that " stones grow, vegetables grow **and** live, and animals grow, live, and feel." There are certain other peculiarities upon which I shall not dwell, **as** they are of little interest to the general reader, who, I doubt **not,** would laugh heartily at the **notion** of mistaking a cabbage for **an oyster, or** this latter **for a common** fire-brick ; yet smile though he may, **I can tell** him **that it is a** matter **of no** small trouble to prove that **a sponge is not a vegetable,** though recent investigations have discovered **in the** sponge a nearer relation to **man than was** dreamed **of when it** served the purpose of completing **our shaving** operations some years since. Physiology has to do with **both plants** and animals, **but** it does not treat of their forms or structure, **these** belonging properly to the sciences of morphology and anatomy, which lend us, as it were, a helping hand now and then; it teaches us how the complex machinery of these beings acts, in a word it deals with their functions. Animals could not continue to exist without plants, for two reasons: firstly, because they would be deprived of food ; secondly, because they would be suffocated; now it is only by calling in the assistance of chemistry **that we can be** made to understand these facts. Chemists have **carefully** examined almost all mineral substances, and have **found that** these, **no matter** how different they may be from **one another,** are composed **of a** certain number of simple materials, which cannot be subdivided **into** any **others;** each simple substance they term an element, and elements may be combined artificially to form minerals. This may **be** difficult **to** appreciate at the outset, but by comparing the elements of the chemist with the letters of the alphabet the difficulty will vanish; thus, **each** word when analysed will be found to consist of a series of elementary parts, or letters, which may again be reconstructed into words. Words in this case represent minerals; letters take the place of elements. Numerous as are the varieties of mineral compounds which lie scattered over the globe, we find that the elements which constitute them are but seventy or thereabouts in number; in fact, the mineral alphabet, so to speak, contains seventy letters; but oh ! how manifold are the words, how complex the idioms, how difficult but withal how sublime the grammar and translation !

I fancy the reader can soon perceive in their full force the advantages of plants to us, which I alluded to above.

Vegetables, like animals, are supplied with earth, water, and air, and from these they manufacture (if that term may be so employed) the food on which animals subsist ; they possess a power of composing, which is denied to animals ; and, to carry our alphabet simile still further, plants hold the same position with regard to animals that a man who can read and write holds with reference to a child who **has** only learned his letters; the first in each instance can form new combinations,—the man new words, the plant new compounds. It is **of** importance to recollect that plants cannot combine any set of elements but only a limited quantity. To illustrate what I have just mentioned, let us imagine a plant to be supplied with earth, air, and water, and it will cause the following changes in these latter:—

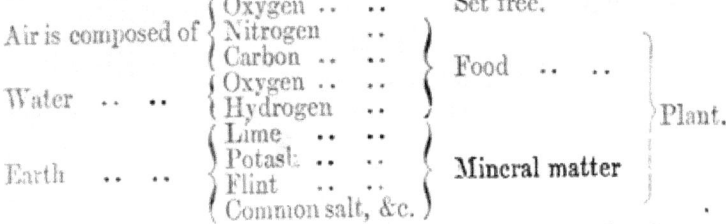

The vegetable has drawn into it certain minerals, has made some of the gases of the air unite with water to form a new compound which may serve us for food, and in doing so has set free a gas called oxygen, which goes back again to the atmosphere; at least by not appropriating it, it has allowed it to remain behind; and this is of the utmost moment, for it is upon oxygen that animals depend — as I shall show hereafter — for the purification of their blood; whilst carbon, which the vegetable world has abstracted, was, when in the air, under a most poisonous form, and, if allowed to remain there, would have proved highly detrimental to animal existence. To complete the balance, I must observe that animals are perpetually vitiating the atmosphere, by pouring into it from their lungs carbon, in the gaseous form I have referred to; thus, what the animal rejects the plant feeds on, and what the plant leaves the animal respires: this the diagram beneath will render explicit:—

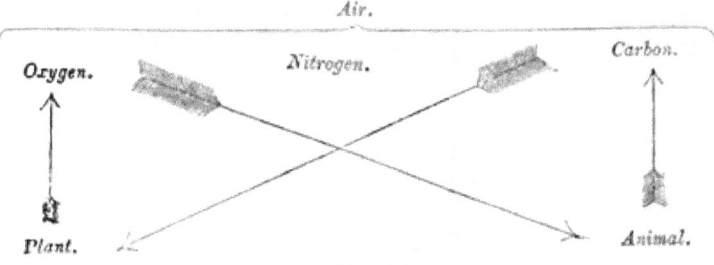

Fig. 1.

The nitrogen plays the part of water in adulterated milk — only dissolves and dilutes. The plant allows the oxygen to ascend from it, whilst it attracts the carbon; the animal allows the carbon to proceed from it, and absorbs the oxygen. This must not be looked on as an accurate and precise explanation, but conveys, in a rough way, an idea of the process, which, were it detailed at full length, and with consideration of recent researches, would be too puzzling and complicated for these pages.

It must not be supposed, from the foregoing remarks, that animals can produce no alteration in vegetable compounds. Though incapable of building up new structures from *elements*, they can bring about changes in food submitted to them; but of this I shall say more in the proper place.

Concerning the application of experimental physics to the study of Physiology, I need only mention one circumstance with which some of my readers may not be conversant, *viz.*, when a thick fluid, contained in a bladder, is immersed in a fluid which is thinner, both fluids pass through the membrane towards each other, the thick one passes out to the thin one, and *vice versâ* (fig. 2A). Thus, if a strong solution (*d*) of sugar be enclosed in a tube (*e*, *f*), whose bottom (*b*, *c*) is an animal membrane, and this be plunged into a vessel of water (*a*), it will be found, after a while, that some of the sugar has passed out, and a portion of the water in; to this permeation the name of *osmose* (*osmos*, a plant) has been applied, the passage outwards being termed *exosmose*, and that towards the inner fluid *endosmose* (fig. 2B).

The laws of mechanics and hydraulics we shall find also applicable to some of the duties performed by parts of the animal body, though it would be out of place to enter upon that portion of the subject here.

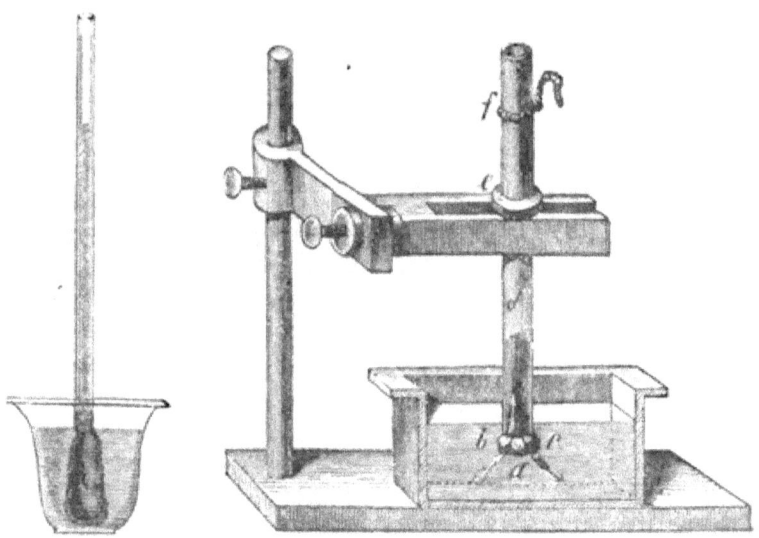

Fig. 2A.   Fig. 2B.

CHAPTER II.

Method of examining Man's Mechanism—Comparison between a Watch and the Human Frame—General plan of the Skeleton—Protective Bones, and Locomotive Bones—Vertebræ—Ideal Vertebra or Model—The Limbs—Resemblances between the Arms and Legs—Soft parts of the Body—Division of the Trunk into two portions: Chest and Belly—Position of the Great Viscera: Lungs, Heart, Liver, &c.

How hard it would be to form any adequate conception of the beauty of the mechanical arrangement observable in a watch, if the various wheels, springs, and appliances which compose it had been previously separated from each other, and given to us altogether in chaotic confusion. What an imperfect geographical knowledge we should possess, did we huddle together the names of the cities, in every nation, without regard to their position and distance from each other. These questions show us the necessity of taking a general survey of the complex piece of mechanism we are about to examine, before we take it bit by bit asunder. When we gain a notion of the geography of the man-world, we shall then examine each particular region; and we can do so with more ease, for in human mechanics, as in the watch, every little structure is in some way dependent on the other; all parts are bound firmly together, and if one portion of the apparatus be removed, the entire combination is seriously affected, and the heart, that human "*balance*," indicates but too clearly that life, the main-spring of the system, has exhausted its hitherto accumulated force.

In the watch the wheels and springs are supported by the framework, and protected by the cover; so in man also have we the framework in the bony skeleton, and the cover in the skin, or, as it is more usually called, the integument; to perfect the analogy, we have only to discover something similar to the movements of the chronometer, and this is no difficulty; all those portions of man's body which have a set office to perform, and which are termed *organs*, are actually the human "*movements*," if we may so designate them.

Nothing can show more fully than the skeleton of man, the necessity for admitting the existence of a Creator; the wonderful design which it exhibits, and the admirable dovetailing of means to end; the grandness of the general plan, and withal, the extraordinary adaptation of every individual portion to the wants and requirements of the being, point to a supreme First Cause, of whose infinite intelligence we can form but a meagre and imperfect notion at best. The human skeleton may be divided into two parts; first, that which is protective, serving to enclose and screen from injury the important organs; second, that which is added to enable the creature to move from place to place, and to provide itself with food. The one I shall term the protective, the other the locomotive division. In treatises

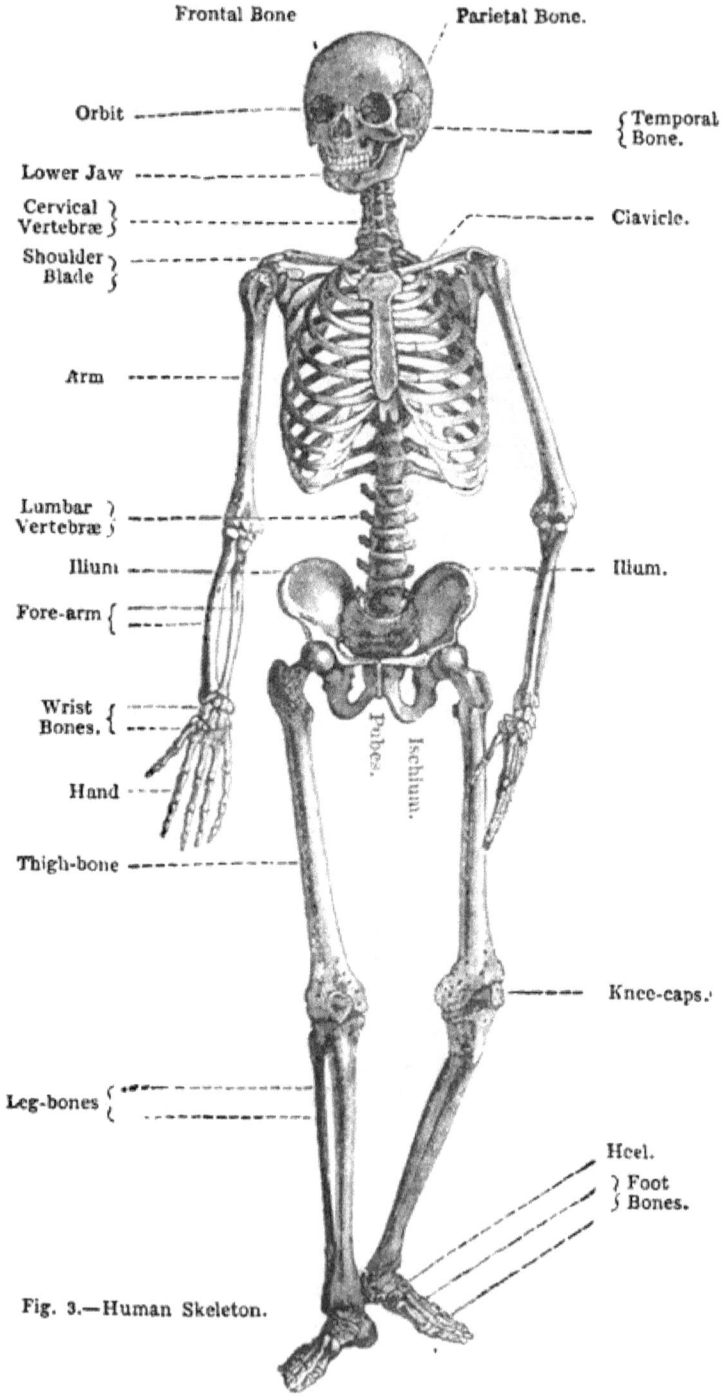

Fig. 3.—Human Skeleton.

IDEAL AND REAL VERTEBRÆ.

on human anatomy it is usual to state that *so many* bones are in the skeleton, and then to begin a dry, technical description of each; this is puzzling enough to the unlucky medical tyro, and, as a matter of course, would be still more so to the general reader, so we shall not follow a scheme of that kind.

If I were to come across a savage, entirely ignorant of the nature and form of the houses we civilized people live in, and I wished to convey to him some idea of the character of our domestic architecture, think you I should give him a lucid conception by commencing in this style: On either side of the hall is a sitting-room, two more in the rear; there is a kitchen, a scullery, and there are four bedrooms, &c.; and then enlarge upon the size of the apartments, their floorings, windows, papering, and so forth? Certainly not. I should begin with the simplest form of house, thus showing him the general plan, and then I should lead him step by step to more complex forms; and so he would have some appreciation of what I meant to convey. Now this is just what I mean to do in giving the "estimate" of man's architecture; we must first glance at the plan, and afterwards *seriatim* take up the details. The protective division of the bony framework is much less complicated than it seems, and, no matter what its length or breadth, is made up of a chain of parts, which are large or small, to suit particular purposes, but which are all constructed on one plan. I do not wish the reader for a moment to suppose that the Creator had that plan before Him "in the beginning." That would be going a little too far; but, to our minds, they are reducible to one kind, just as we may say all lighthouses are built upon a common plan—viz., a means of producing a light-signal, and a means of erecting this signal so that it may be seen. The common fire beacon of remote seaside villages, and the complex Eddystone, have for us certain things in common, so that we picture to our minds an ideal lighthouse, and modify it to suit our pleasure, without troubling ourselves to consider whether the architects possessed *our* plan or not. In a loose way, we may look upon the protective skeleton as being constituted of the Back-bone, Chest, and Skull; but when we come to consider each of these in its turn, we shall find, as I mentioned, that it is composed of a chain of bones, resembling each other strongly, and which I shall call vertebræ (fig. 4). These are—if I may be excused the comparison—the bricks of the edifice, and, like these latter, have a bit lopped off now and then to suit a special end.

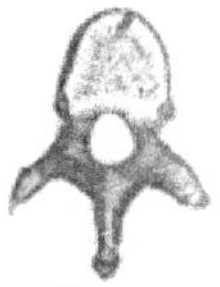

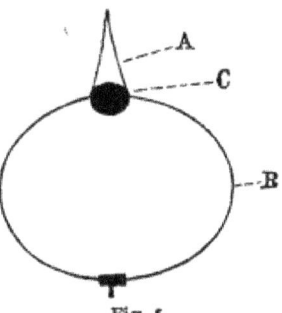

Fig. 4.

Fig. 5.

Before we cast our eye over the different vertebræ, let us look at our model (fig. 5).

It consists of a solid centre, or body (*c*), to which is conjoined, on the upper part, a triangular arch (*a*), prolonged into a spine, and from whose lower surface spring two stout ribs, which unite with a single piece of bone in front. This is our typical vertebra; and as I go on, the reader will observe how simple and beautiful the arrangement of the skeleton appears, when we, as it were, unfold its mysteries by the application of this key. We begin with the chest. This bony cage, which during life contains the heart and lungs, is really nothing more than twelve of these vertebræ (*d*, fig. 6) piled up one on the other, the upper ones having smaller ribs than the lower, and a few of the last having their ribs not united in front. The breast bone is only the sum of the front bones of the vertebræ, which have become soldered together. How is this a protective case? The triangular arches, locked one in the other, form a long canal, which contains the "spinal marrow," a great nervous organ, and the ribs cover in the delicate apparatus of the lungs and heart, which, if exposed to external pressure, would soon cease to carry on the vital processes. Between the chest and hips we have a column of bone which supports the former, and rests upon another that we shall speak of presently. This column is composed of five vertebræ (*l*), which are like our model, with this difference: their ribs and front pieces are not bony, but of a sinewy, or flexible description. It would be singularly unfortunate were there osseous ribs in this situation, for within lie the stomach and intestines; and oh! what would become of the Aldermanic race were not this region capable of distension? Nature, foreseeing the future advancement of the gastronomic art, has wisely provided against the horrible contingencies that *might* arise, if a man's belly were boxed in bones. Beneath the pillar of vertebræ I have been just describing lies another, which, however, has its component parts so fused together as to be inseparable. This is firmly wedged between the bones of the hip, and is the foundation-stone of the whole spine. It is called the sacrum (*s*), from the notion which prevailed formerly, that when at the last day all the other bones had crumbled into dust, from it the whole skeleton would be reconstructed. Its ribs do not complete the circle and unite in front, but are merely flattened pieces, which connect it to the hips on each side. The tail, or coccyx (*cx*), as it is termed, from being supposed to resemble the bill of a cuckoo, consists of four little bones placed end to end, and continuous with the sacrum; in these our model vertebra has been deprived of all its appendages, and nothing but the central body remains. I trust the reader will not be surprised to learn that he does possess a tail, for that such he has is unquestionable; and were the bones which form it drawn out a little, we should be then provided with a decided but I fear not very dignified caudal extremity.

Fig. 6.

We started at the chest, because that locality furnished the best illustration of our model; we must now proceed to deal with the other regions. The neck is formed by a row of seven vertebræ (*c*), resting upon those of the chest. These have their central parts; and back arches containing the spinal marrow; but where are their ribs? They do not come round to the front? No! The reader will be astonished to learn that they twist backwards on each side, and in this way make a series of little bony rings, through which run certain blood-vessels to the brain. Ay, but why are they not arranged as they are in the chest? Simply because, in the first place, they have no organs near them which require *great* protection; and in the second, let me ask you, Sir, Madame, or Miss, how you could possibly give your friend that familiar nod, or indulge in that quiet confabulation, or get down that dear blue pill or bolus, if there was a great bony collar round your throat, just like that of **a crocodile?** The last portion of our protective framework is the **skull.** This, complex though it seems at first sight, may be reduced **to three** vertebræ, one of whose ribs is represented by the lower jaw. It is fixed on the top of the back-bone, of which it is an expansion; and just as the back-bone contains the spinal marrow, so does the skull **contain the brain,** which is the continuation of that marrow. It would be too perplexing to the reader to wade through this subject further, so I shall now say a few words about—

The Locomotive Skeleton.—This is constituted of two pairs of extremities, somewhat similar in form, and attached to the protective framework by intervening bony girdles, which, though apparently very different from each other, have very many characters in common.

These girdles may be distinguished by the terms "upper" and "lower"—the first sustaining the arms, the second the thighs and legs. Each is composed of six pieces, or three pairs; those in the upper one are, the shoulder-blade, the collar-bone, and the coracoid* of each side; and in the lower one, the hip bones. To the reader there may be some slight difficulty in grasping this idea, for the hips are, in the adult, represented by two solid buttresses of osseous material, but in the infant, the hip on each side is composed of three distinct portions—the ilium, ischium, and pubes, which in after life become so blended together as to leave the mark of union obscure. We have, then, three pairs of bones in each girdle, which represent each other, as shown in the subjacent table:—

Upper Girdle.	Lower Girdle.
2 Blade bones	2 Ilia.
2 Collar ditto	2 Pubes.
2 Coracoids	2 Ischia.

In the upper belt the three components unite to form a socket for the head of the arm, and similarly in the lower one the point of union of the three bones is situate in the hollow excavated for the ball of the thigh. The two ilia may be looked on as the blade-bones of the hips. The shoulder blade moves freely on the back of the ribs, because it is necessary that the arm should possess great freedom of motion. Now, the blade-bone of the hip is attached to the stunted ribs of the

* So called from being supposed to have the form of a raven's (*Korax*) beak.

lower vertebræ; but in order to give firm support to the entire trunk, it is strongly soldered to these ribs, and does not glide easily over them, as in the shoulder. The extremities are made up of similar parts, arranged on the same plan. There are two pairs, one for each girdle; the upper being for the purpose of prehension or grasping, the lower for locomotion. The prehensile extremities consist each of a long shaft, working by a ball-and-socket joint in the shoulder, two smaller bones lying side by side, and hinging upon the end of the first, two rows of small squarish bones at the ends of these, and finally the hand. The long shaft is the arm, the two following make the fore-arm, and the double row of small bones constitutes the wrist (*vide* fig. 3). In the thigh, leg, and foot, the conformation of parts is much the same, but can be more readily conceived by contrasting the two limbs, and by glancing at the table beneath:—

LOCOMOTIVE EXTREMITY.		PREHENSILE EXTREMITY.	
Thigh	.. .. 1 bone.	Arm	.. .. 1 bone.
Leg	.. .. 2 bones.	Fore-arm	.. 2 bones.
Heel	.. .. 7* bones.	Wrist	.. .. 8 bones.
Foot.		Hand.	

It is to be regretted, that space will not permit me to enter fully into the wonderful evidences of design exhibited by the various bones, joints, and mechanical contrivances of the skeleton; but it would be quite foreign to the scope of a writer on pure physiology to enter into more details; and indeed I feel that I have already trespassed too much upon the province of the human anatomist in entering so far into the subject.

Up to this we have been considering the hard parts of man; we must now take a peep at his softer ones. These embrace two groups— that which helps to carry on the processes common to animals and plants, the vegetal; and that peculiar to the former, the animal group. The vegetal division embraces the heart, lungs, liver, kidneys, intestines, stomach, &c.; the animal portion includes the brain, spinal marrow, nerves, muscles (flesh), and tendons. The first set is divided between the chest and belly, but in both positions is guarded by the ribs, and has the osseous column of the back-bone behind it. The chest and belly are not continuous, inasmuch as a great oblique mass of flesh, or muscle (diaphragm), which is adherent all round to the last ribs, shuts off one from the other, partition fashion. Above this septum, and within the bony ribs, lie the lungs, having the heart and great bloodvessels between them; and below are placed the liver, stomach, kidneys, bladder, sweetbread, and intestines, as shown in the adjoining cut (fig. 7). The organs of the so-called animal class are of two kinds —the nervous and muscular. The nervous system, of which a more detailed account shall appear elsewhere, comprises the brain, spinal marrow, and nerves; the two first are enclosed in the canal formed

* Though there are but seven bones in the adult heel, yet the true heelbone is the result of the fusion of two distinct portions.

THE SOFT PARTS. 11

by the vertebræ, which, piled up one above the other, compose the back-bone; and, by expanding at the top of this latter, the skull also.

The muscles clothe the entire skeleton, and are either long bands or triangular pieces of flesh, arising from one set of bones and inserted into another, whose actions we shall study in a future chapter. Finally, the integument, or skin, is a more or less dense covering which sheathes the entire body and contains a vast series of glands, which form the perspiration fluid.

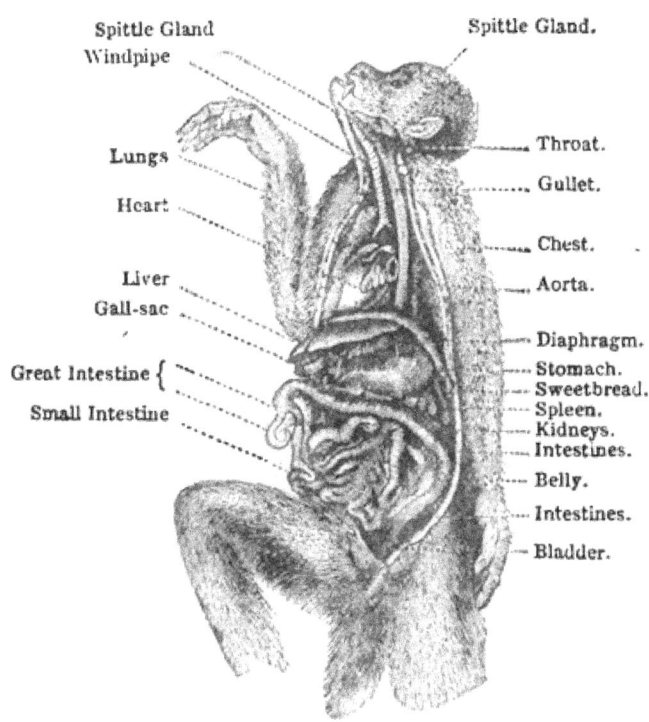

Fig. 7.—Section of the Chest and Belly of an Ape, showing the positions of the different organs situate in these regions. The parts are arranged in an exactly similar manner in the human body.

CHAPTER III.

Life—Opinions as to the Nature of Life—Advantage of not defining Life—Discoveries of Berthelot—Force—Correlation of the Physical Forces—Views of Grove—Concealment of Light and Heat by Plants—Organization—Action of Platinum on Oxygen and Hydrogen, and of decaying Meat on Sugar—Development of the Lower Plants—Indestructibility of Force and of Matter—Death—Importance of knowing when Death is present—Case of Drowning—Meaning of the word Physiology—School of Materialists and School of Principle-ists—Sir E. B. Lytton's Novel, "A Strange Story."

WHAT is life? This is a question which has puzzled almost every physiologist from the time of Aristotle downwards; and which, it seems, we are not likely to get a reply to for some considerable period. There is hardly a work on physiology extant but what contains some peculiar proposition relative to the nature of life; and I think we may pretty safely conclude, that by far the greater number of the definitions which have been framed, are little more than general expressions of what we *understand* by the term, without any attempt at an explanation of the phenomena it involves. Thus, we find one set of *savans* asserting that life is the operation of a vital principle; another, that it is due to a special organic force; a third, that it is the total of the processes which resist death; and a fourth, that it is the sum of the actions performed by a living body. In fact, life is life; and there's an end to it. The reader must not expect us to display our ignorance. We are philosophers. The public will be content with some long-winded explanation, which it repeats parrot-like, or as a child does its prayers. Hence it follows that so many false statements have been made. There never has been an induction on strictly logical principles concerning the nature of life; for the simple reason that we are as yet not sufficiently acquainted with the facts. All knowledge consists in facts and ideas; and in a science such as physiology, no advance can be made so long as men form the ideas first, and then distort the facts to suit their darling speculations. It is much wiser to confess our general ignorance of the matter, just stating what we do know, and leaving it to the future investigations of others to complete our labours. If we were to commence the erection of a building, and discovered after a while our utter inability to complete it, would it not be better to wait patiently (just preserving the fruit of our work from the influence of the weather), till an architect arrived, who was sufficiently skilled to terminate the operations we had undertaken, than to continue the process ourselves, and so not only construct something which must tumble to pieces in the course of a few months, but to display our own bungling ingenuity to all who had the capacity to perceive it. Let it here be admitted, then, that we do not yet know what life is; but, furthermore, let it also be said that we are making rapid strides towards an elucidation of the mystery, and that it is not impossible that, before the end of the present century, we may possess very lucid ideas regarding the true cause of the different vital phenomena. Physiology is in a great measure dependent for its advancement upon the progress of the other natural sciences, and of organic chemistry; and as long as such extraordinary discoveries as those of Berthelot continue to be recorded,* so long will the science of life remain oscillating.

One of the most distinguished investigators in the physical science world—Mr. Grove—has for some years been enunciating the doctrine, that but one force exists in the inanimate world; that is to say, that light, heat, electricity, magnetism, gravity, &c., are not distinct forces, but one and the same, which, as it were, shows itself in different ways, under different circumstances, as it passes from our observation, but which is never lost. Nothing can be more sublime than this generalization, or more calculated to ennoble our sentiments of the Creator. Matter is indestructible, so also is force; and the two, mutually co-operating, originate all those phenomena of which our senses, carrying to the mind ideas, enable us to unravel and understand. Let us endeavour, in a rough way, to illustrate this grand law of Grove's. A seed is placed in the earth, it germinates, and after successive years of exposure to the influence of this one force (as heat and light), it draws mineral matters from the earth, and woody material from the atmosphere—in fine, becomes a lordly tree; owing to a series of geological changes it is converted into coal, this coal is thrown into the grate of a locomotive, it is then resolved into the separate elements of which it was originally composed, and light and heat are evolved; the latter operating upon the water in the boiler causes it to expand, producing steam; steam operates upon the machinery, giving rise to mechanical force; this urges forward the engine, overcoming the resistances of the atmosphere and friction, and so the original force glides out of our sight, to return again in some one or other of the foregoing conditions. The schematic arrangement beneath may assist in explaining what has been described above:—

Germ { Mineral matter, water, carbonic acid, ammonia } = tree = coal { Mineral matter, water, carbonic acid & ammonia } when burnt.

Plant { conceals the force which applied to it during life as .... } Light and heat.

Coal { sets free the force which had been concealed by plant as .... } Light and heat.

This explanation must not be regarded as strictly accurate in detail; but only such as may serve to give the reader a notion of the actual

* Berthelot has within the past twelve months discovered a process, by means of which he has united two gases, oxygen and hydrogen, in such a progressive manner as to produce alcohol (spirit of wine).

"correlation of the physical forces." Many of the functions performed by the human organs can be explained by reference to the ordinary laws of chemistry and physics; but there are certain other processes which are as yet inexplicable, coming under the denomination of *organization*. Thus, the reader may see that our real knowledge of life is that of about a moiety of the entire phenomena, and for those operations of the living body which we are not yet in a position to understand, we have coined the expression organization. The following history of the early life of one of the lowest forms of vegetables will illustrate this more fully.

In its adult state this plant is a small vesicle, or bladder, very transparent, filled with fluid matter, and small granular particles; in dying it bursts, and sets free the molecules, each of which commences converting itself into a structure the same as the parent, and this act is thus performed Being surrounded by an atmosphere containing watery vapour, carbonic acid gas, and ammonia, it so affects these latter, as to cause a breaking up of some, and a union of others, till it develops a material similar to the white of egg; when a sufficient quantity of this substance has been collected around the granule, the outer layer undergoes peculiar changes, by virtue of which it assumes the features of a membrane enclosing the entire mass; next, the granule in the centre draws *into* it some of the matter which it has developed, and causing it to partake of its own characters, then splits up into several portions, after which the membrane bursts, and so on, as before.

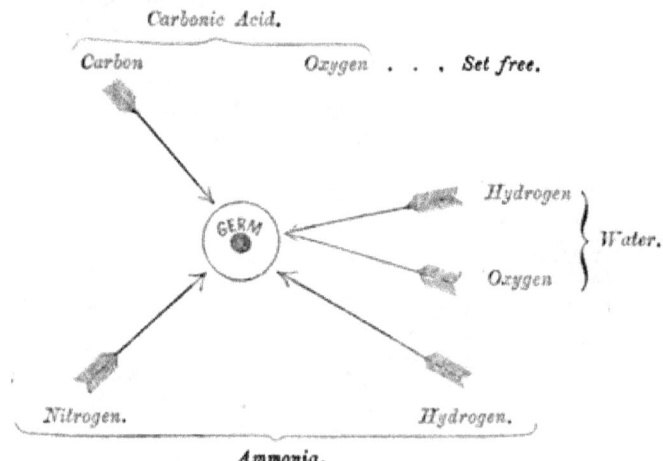

Fig. 8.

The first of these changes, which I have thus endeavoured to delineate diagramatically, is most probably a purely chemical one; for we know that platinum, in a state of very fine powder, will cause the union of oxygen and hydrogen gases and the formation of water;

and a piece of animal flesh thrown into a solution of sugar will break that compound up into its constituent parts. The second and third come under the head of organization, that is to say, we believe they are performed in obedience to the laws of organization, or, in other words, we know nothing whatever about them. It might be, that the conversion of the outer portion of the jelly-like substance into membrane is partly due to chemical change, and in some measure also to a change of temperature, whilst the division and growth of the contained particles may be a process akin to that of crystallization; but be they what they may, we are not yet sufficiently acquainted with these organization-changes to lay down any law respecting them.

If the reader can comprehend the remarks I have now made, he will have as clear a notion of the nature of life as the first physiologist in the world; our information may be tersely written as follows. "Life is the mutual co-operation of the physical forces (or *force*) and organized matter." Life is a condition of perpetual change, and the greater the vital activity, the greater will be the extent of this change. Every action of a living being, every movement, every thought is accompanied by the waste of a definite amount of structure, just as the larger the surface of the wick within a lamp, the brighter and more luminous the flame, but proportionally the more extensive the exhaustion of the combustible pabulum, or supply. During life certain materials robbed from the inorganic world (having been retained, *pro tempore*, within the body in disguise) pass from their living prison, and are restored to the source from whence they were primarily derived. How sublime is this scheme, and how significant of the truth of that maxim which says, "All nature moves in circles." "Dust thou art, and unto dust thou shalt return." Assuredly this is but too true; and so is the remark of Hamlet:—

> Imperial Cæsar dead, and turned to clay,
> Might stop a hole to keep the winds away.

Death, the antithesis of life, we cannot fully interpret till we increase our knowledge of the latter. If we suppose life to result from the exhibition of "force" operating on organized bodies, and then conceive that this force, not supplied with the conditions favourable to the production of vital action, has degenerated into chemical decomposition of the organism, we shall have *some* idea of death. It is frequently stated in text-books on physiology that death is of two kinds: first, that in which all the functions of the body cease, but where the latter does not become putrid; secondly, when putrescence makes its appearance. The former description of death has the term somatic (*soma*, the body) applied to it, whilst the other is called molecular. It seems unjust to call that state of the body death, where vitality has ceased *only so far as we are capable of observing it*, because, though unperceived by us, life may linger still, the last faint flicker of the lamp of existence may still be there, the flame of life may not yet be extinguished, as shown in many cases of recovery from the effects of drowning. One instance is remarkable as bearing especially upon this question. A man was removed from the water in which he had been immersed, and was found, to all appearance, perfectly dead. The usual appliances were tried unsuccessfully; he seemed completely asphyxiated; the skin was cold and clammy, the heart had ceased to beat, there was no breathing,—in fact, there seemed not the slightest reason to hope for restoration to life. Some good friends, less impatient and impetuous than others, determined that they would leave no means untried to resuscitate him, and with this object caused him to be placed in a warm bed, placed hot jars to the feet and stomach, and continued to apply friction in the most praiseworthy manner for *nineteen* hours; at the end of which period, to their intense delight and wonder, the symptoms of returning consciousness made themselves manifest. In this case there was *every* indication of death having taken place, and the modern physiologist would inform you that "somatic death" held his sway. For my own part, I think the use of these designations is more calculated to confuse than anything else. As we are yet most ignorant of the character of death, we ought to employ the expression in its correct acceptation, *and only consider a man dead when the first stage of decomposition has made its appearance.* Having so far treated of the distinction between life and death in the most cursory and outlinear manner possible, it behoves us to inquire into the true limits of the subject we have taken in hand—physiology. This is the science of life; that branch of study connected with the phenomena of living things, or which teaches us how the apparatus of animals and vegetables work. The name is derived from two Greek words—*phusis*, which means the supreme animating principle, and *logos*, which signifies a discourse.

Dear me! what a strange and complicated piece of machinery! How does it act? Who has not at one period or other of his or her life made these remarks? I shall fancy them addressed to myself, and shall in the following pages undertake the task of conducting the reader from the basement to the upper story of the human factory, and of describing in as brief a way as is consistent with clearness the various processes by which the crude material, in the shape of food, is eventually transformed into the delicate and complex fabrics of the human body.

Before terminating this chapter let me commend to the reader who would learn in the *most pleasing* mode the difference between the two schools of physiologists—materialists and principle-ists—the recent production of Sir E. B. Lytton, entitled "A Strange Story." Nothing could be more lucidly, attractively, or correctly stated than the cases of the two sides by this eminent writer; and although I cannot concur in the opinions* of the distinguished novelist upon certain questions, still, I confess to having been amply repaid for the time spent in the *study* (*it must not be read lightly*) of this most interesting and philosophic work.

---

* If indeed an author be responsible for the opinions expressed by any of the characters introduced into his works.

CHAPTER IV.

Tissues—What is a Tissue?—Controversy among Physiologists—A Cell—The Cell Theory—Development—How a Sinew is formed.

THERE is still another branch of our subject which, though somewhat introductory, we have not yet touched on. I fear some may think these details too dry and technical, but it must be remembered that all beginnings are more or less uninteresting, as the fair reader, who has invariably skipped over the preface to Scott's stories, may perchance be aware. "I come to speak" on the subject of man's tissues. You will perhaps inquire, What is a tissue? It is one of the materials of which the frame is composed, and which *seems* to result from the weaving together of certain lesser parts called fibres or cells. Every portion of the human tabernacle is tissue of some kind or other. It builds up muscle, and is termed muscular tissue, or nerve, hence nervous tissue, or bone, from which bony tissue, is employed to form an organ which extracts substances from the blood—a gland—and is then termed glandular tissue, and so on. Now, as to the minute structure which these different fabrics present when examined with the highest power of the microscope, an awful battle is taking place among physiologists at the present day. The conflict is not confined to Britain, but rages over the Continent also, and the first men in this department of science have put forward their opinions. But this is not all; another section is at war concerning the infant features of the tissues and regarding their genealogy; and as the questions referred to excite considerable interest among scientific men, a short *résumé* of what has been proved on each side may not be unacceptable.

It is necessary in the first place to state that the controversy turns on this point: Is the body composed of little vesicles or cells? And here I must premise my outline by giving the following definition of a cell:—it is a minute bladder, closed all round; its outside resembles gelatine, that is to say it is perfectly transparent, and has no (apparent) structure; inside there is liquid, and floating in this a small granular rough mass called a nucleus; withal, it is so diminutive as to be seen only by the help of a good microscope. A German anatomist wrote a book some years ago to prove that every tissue in the human body was made up of these cells, and, I suppose, satisfied *himself* of the truth of his hypothesis; it was a very nice idea, and "took" exceedingly well for some years, having been supported by many of the philosophers in this country and Germany; but "every dog has his day," and so with the poor "cell theory," for after a short reign a most Socratic Brutus slaughtered it unmercifully, but justly; if you would ask me the name of this tyrant slayer, I should not tell you: he is one whose claims to distinction have not yet received the recognition they merit so well; he is an Englishman, and a naturalist and one who will hereafter be regarded, in comparative anatomy, as the Bacon of the nineteenth century. *Verbum sapientibus,* he bore "a banner with this strange device" — Development. Ay, but what is development? Ah me! What a nuisance it is to have these horrid long words turning up so often. Lend us a hand then, or we shall stumble. I cannot define development more tersely than it has been done by a recent writer, who says, "it is the process by which a being or organism is brought from what it *was* to what it *is.*" To return, then. This gentleman said, If you want to know what a structure *is,* first learn what it *was;* and pursuing this view he worked indefatigably at the development of portions of the body, and showed, *I* think, conclusively, that at all events *all* tissues are neither cells in the adult nor in their infant state; so the German theory, as universally applicable, we may regard as being now (to use a slang expression) absquatulated.

As regards the earliest condition of all tissues, there is yet much to be learned. A most interesting theory has been put forward within the present year, but inasmuch as it may be regarded as *sub judice,* I shall not enter into it here. The doctrine I have been accustomed to teach, is one that was promulgated some years since, and which I regret to find has not been treated with the consideration it deserves. In a few words it is this. All structures in the body, no matter what; be they nerve, muscle, skin, or bone, start in the same way, and from the same point, on their journey to *what-they-will-be;* one may take this line, another that; one may travel by *express,* another *parliamentary;* each may arrive at a different terminus, but *all* commence their adventures under the same form. Let us, for example, take a bit of a sinew, and watch its progress: first (A, fig. 9) it consists of a jelly-like transparent substance, through which (like raisins in plum pudding) are scattered numbers of little granular particles, imperceptible to the naked eye. After a while we find the jelly substance expanding in such a way as to have a small cavity round each granule (B). These cavities increase in size and run into each other, so that we have now the sinew composed of solid bands of jelly, and hollow bands containing granules as in C.

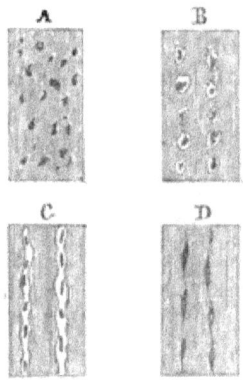

Fig. 9.

Next, we find the walls of the hollow bands closing in on each other, and converting themselves into fibres, whilst the jelly hardens and assumes the appearance of a number of flattened threads (D), and the perfect sinew is formed.

This description must suffice for all, and if it is stated so that the reader can apprehend the various stages of the development, a great step will have been made. The cavity which the intervening soft material leaves round the granule is what the German anatomist mistook for a cell; of course, in the *ordinary* sense it is a cell, but if employed in this acceptation it would not be of any use in carrying out his views, and in the strictly physiological meaning it is *no cell at all.*

The account above given will answer in a rough way for all tissues; but if it be asked, Why does one become flesh and another bone? I answer. We cannot tell. Why does one mineral assume one particular geometrical form, and another another? Did the narrow limits of this volume allow me, I should here enter minutely into the description of the microscopic characters of bone, flesh, nerve, and such like; but it would occupy too much space, and would, I dare believe, be tedious to the general reader; therefore, I shall now pass on to the more truly physiological branch of the subject, merely observing that for the minuter features of the tissues, the following books may be consulted with advantage: Carpenter on "The Microscope and its Revelations," Todd and Bowman's "Physiological Anatomy," Quain and Sharpey's "Elements of Anatomy," and Kölliker's "Manual of Human Histology."

CHAPTER V.

Food—Classification of Aliments—Alcohol—Opinions of Teetotallers—Effect of Spirits upon the Body—Combustion and Decomposition of Alcohol—Duchek's Experiences—Composition of Dublin Porter—Value of German Beer—Views of Liebig—Importance of Sugar as a form of Food—Blubber and the Esquimaux—Nutritious Properties of Flesh—Cheese—Materials necessary for the formation of Bones—Use of Salt—Constituents of Milk—Tissue-repairing Power of Tea and Coffee—Gelatine and Soups—Are Broths and Jellies nutritious?—What is a Poison—Popular Idea of the Action of Mercury—Strychnine the Food of a Bird—Bread made of Flour and Clay—No animal can live upon *one* form of Food—Experiments of Magendie—Death of Dr. Stark—Cause of Hunger—Thirst and Satiety—Is the habit of smoking Tobacco injurious?—Common Sense *versus* False Philosophy.

IN that self-repairing machine, the animal body, the raw material which is woven into the several tissues is termed Food. One would think it could not be so very difficult to know what food is, but it is not so easy a matter after all, and food has for some time past been a *bone* of contention to physiologists.

We may say that food is a substance or combination of substances which supplies to the body some material or materials of which it is composed. It is of the following kinds:—

* Except possibly to the misguided Deutschlander.

FOOD
- Liquid
  - Water (1).
  - Alcohol (2).
  - Beer, wine, &c. (3).
- Solid..
  - Not mineral
    - Fats, starch, sugar, gums (4).
    - Flesh, eggs, gluten, &c. (5).
  - Mineral .. Common and other salts (6).

Heretofore, it has been customary to classify the various forms of food under two heads,—those of *nutritious*, and heat-producing. This method, however, was not based upon a perfect knowledge of the parts which each class of materials played in the body, and hence I shall not adopt it. Teetotalers have long objected to the view which holds that alcohol is food, asserting that it is a poison, and shocking the minds of all rational men by advocating the introduction of a "Maine liquor law." *Media tutissimus ibis*. We need not run into either extreme; we may surely "send round the bowl" without reducing ourselves to the condition of beasts; and if a portion of the population will be brutes, the rest should not be deprived of reasonable indulgence in alcoholic drinks. By parity of reasoning, because a certain percentage of the British people annually select carving-knives as a medium of conveyance to *another* world, the remainder should at once be obliged to employ "chop-sticks" for the future. No, no, dear reader, be assured that teetotalism, as also phrenology, mesmerism, electro-biology, homœopathy, and such like, will, in the course of another century (when *our* bodily atoms are transformed into cabbage-plants), be looked on as a few of the eccentricities of this enlightened age.

Before I attempt to show the use of alcohol, it must be understood that the waste of the body which food repairs is not entirely without an end or purpose; the refuse or effete matter is absolutely burned out of the system, and such burning produces the heat which animals could not exist without. If some compound be introduced into the body which will burn of itself, the tissues do not then undergo *so much* destruction, because there is no special need of their consumption; now, this is exactly what takes place when spirits, whether in the shape of wine, whisky, brandy, rum, or beer, is taken into the stomach, and from this we learn that, though not acting directly, they act *indirectly* as food. If the alcohol is burnt in the spirit-lamp (which is the series of minute bloodvessels) of the animal economy, it must be altered in character, because it must unite with the oxygen of the air which has been carried from the lungs. The question turns on this point; and as recent investigations of teetotalist philosophers go to prove that the alcohol passes unchanged from the skin, lungs, and kidneys, their researches deserve some consideration:—

1st. Supposing their experiments to have been fairly and accurately conducted, they only prove that certain minute portions of the alcohol passed out in a pure form, and these might be accounted for by the fact, that enormously large doses had been administered; and they have failed to prove that *all* the alcohol was expelled unchanged.

2d. Supposing certain portions *did* pass out unaltered, this would not prove that it was not food, because as certainly does water (which *is* food, and part of which *is* decomposed in the body), pass out in its pure form also.

3d. Duchek and many others have found it in the *altered* state in the different tissues.

That spirits, in the form of whisky or brandy, do maintain the animal heat, I fancy no one who has ever enjoyed (?) a day's sporting, or spent a week investigating the geology of a bleak mountain district, can for a moment deny. I myself have experienced the delightful glow of warmth which steals over the body after a long day's exposure to hills and hunger, on the application of the lips to a well-charged flask of cognac; and I fear I am too much of a sceptic to be readily convinced that the pleasing sensation was due to the *irritating and poisonous* effects of the "eau de vie."

As bearing on the alcohol question, I must say a few words for our beer. I have often heard people decry beer as a beverage "fit only for besotted boors;" but when I consider the enormous proportion of beer-drinkers in the population of Europe, and the composition of the venerable John Barleycorn himself, I feel sufficiently philosophic to adopt the common-sense opinion—that it is no sin to be a Malt-ese. Porter and beer we may class together, for the effects of one are pretty nearly the same as those of the other, and hence an analysis of the first will give us a clue to their common properties.

In 1,000 parts, by measure, of porter there are of—

Proof spirit	89·6
Vinegar	3·6
Grape sugar	1·72
Albumen	8·272
Phosphates	0·851
Common salt	0·448

A glance at the above table must be sufficient to convince even the most sceptical of the highly nutritious quality of this form of food. It is true that Liebig believes that a man who has been drinking several pints of beer per day does not receive more nutrition at the end of the year than he who has just eaten a 4 lb. loaf; but any one who has travelled in Germany will admit the existence of a very well marked distinction between the "good Bairisch bier" of that country and London porter. I must say that to my taste the Bairisch and Christiania beer drunk in Deutschland seems more like the washings of an English brewery than anything else; and if the philosophic Baron indulged in a twelvemonth's course of these fluids, with a view to testing the truth of his own researches, he certainly deserves well of his country.

In fine, I may observe, as an argument in support of the use of spirituous liquids, that there is hardly a nation on the face of the earth, no matter how aboriginal, but what possesses its own intoxicating drink; and the Scripture narrative of Noah's little peccadilloes I need hardly remind the reader of. Schiller, in his beautiful poem, "The Four Elements," thus describes the addition of the various constituents of the emblem of hospitableness and conviviality:—

> Four elements joined in an emulous strife,
> Build up the world, and constitute life.
> First from the citron the starry juice pour;
> Acid to life is the innermost core.
>
> Now let the sugar the bitter one meet;
> And the strength of the acid be tamed with the sweet;
> Bright let the water flow into the bowl;
> For water in calmness encircles the whole.
>
> Next shed the drops of the spirit within;
> Life but its life from the spirit can win.
> Drain quick, no restoring when cool can it bring;
> The wave has but virtue drunk hot from the spring.

So much for the subject of alcohol. Baron Liebig, who may be regarded as the first chemist of this century, thinks that all foods, such as fats, oils, starch, &c., are only employed to maintain the animal heat, and are usually burnt up in the blood-vessels. We can hardly doubt that combustion of these materials does take place (without absolute flame); but I think the great German philosopher goes one step *too* far when he says that is their *only* purpose. Fat is as important a constituent of the tissues as white-of-egg, or any other substance; and when it is not taken in some shape or form, the body becomes emaciated: this we know from two circumstances; first, every portion, to the smallest microscopic particle, of the soft parts contains fat; secondly, when a person is unable to eat fatty materials, and that dread disease consumption has fastened itself upon the lungs, it is only by the use of such medicines as will enable him or her to digest fats, that we can ever *hope* for a cure. In connection with this idea of the cause of consumption, a most remarkable fact has been noticed, *viz.*, if in a family of ten children there is one who for years has exhibited an aversion to fat and butter, it is more than probable, nay, almost certain, that should any member of that family be, in after years, attacked by consumption, it will be the one who has shown the peculiar distaste referred to. But bounteous nature has provided an alternative, which, if adopted, often serves to eradicate the germs of scrofula. It is this: sugar is capable of being converted into fat, and children, who abominate the notion of fatty or buttery food, absolutely crave for sugar; and here, *en passant*, a word to economical (?) mothers. Never forbid your children the use of sugar, forsooth, because Dr. What-d'ye-call-'em says 'twill ruin their teeth, or prevent digestion. Don't deny the innocent darlings the pleasure of spending their little pocket-money in lollipops; for be assured that in so doing you are only helping too abundant conditions to sow the germs of what I believe every family possesses in some measure—*scrofula*. Let us, then, lift our hats to the pastrycooks, for, mayhap, they are labouring, unwittingly, to save us from more ill-health and unhappiness than we imagine.

It must not be thought that sugar will suffice for the wants of the animal economy; but that it is converted into fat we know from this fact; bees when fed upon it will produce wax, and wax contains much oleaginous or fatty matter. However, experiments made in France to determine how long bees could continue to exist on sugar alone, show that, after a lapse of time, the insects cease to work, and if not given honey they do not survive. That fat and oils maintain the temperature of the body cannot be doubted, although it is equally true that all the tissues, in giving rise to "sewage," or effete materials, do the same. It is said of the Esquimaux, that one of them can devour as much as 20lb. of blubber, with the greatest gusto, in a single day. Every one knows of the proverbial predilection of the Russians for grease. I have heard the master of a Hamburgh whaler state, that on one occasion, having sent one of his crew (a Russian) down to the hold to see after some cases of tallow candles, he observed the fellow, on his return to the deck, spit from his mouth a whitish cottony mass, which on examination proved to be the wicks of some two or three candles. Moreover, if I mistake not, some one has asserted that, on the approach of cold weather, the London butter retailers increase the price per pound, knowing that a greater demand will be made (as it were instinctively) for this kind of food.

Flesh or meat, which is the next variety of food, is considered to be the most nutritious; but as nutrition itself is a *very* relative term, it is difficult to admit the truth of this assertion. That the substances of which it is composed are capable of building up most of the animal structures is true enough; but on these alone man cannot subsist. When flesh or muscle has been exposed to water for a considerable time, it is converted into a liquid resembling, in every particular, white-of-egg; indeed, it must be reduced to this condition before it is of service in repairing the bodily waste; and from this there seems very good reason for the view of its high nutritive properties. To show this it is only necessary to refer to the egg, and the young fowl. What is an egg? It is an oval mass of organic and mineral matter, consisting of the white (albumen), fat, sulphur, and salts. Now, as out of these the perfect fowl, with flesh, blood, bone, sinew, skin, brain, and feathers is constructed, we have here adequate evidence to prove the great nutritious qualities of the *entire* mass. Vegetable food, some folk would say, should not be classed under the head of "flesh;" this is, however, an error, and the old division of aliments into fish, flesh, and fowl is also no longer tenable. Besides the albumen which flesh is formed of, there is in the body another substance with which every one is familiar, under the name of cheese. It is found in milk, and differs from albumen in this, that if you add some vinegar to a liquid containing cheese in solution, it falls to the bottom of the vessel as curds; in fact, this is the theory of cheese-making. Vegetables contain two substances identical with albumen and cheese, and which may be extracted by peculiar methods. The first is found in all kinds of corn-grains; the second in seeds of the **pea and** bean kind. This is known even to the Chinese, who manufacture from beans a sort of cheese very much resembling many of our British kinds.

As "there is never smoke without fire," so do we see there is some truth in the view of vegetarianists. Many would laugh if told that bone-dust was food; but let them reflect for a moment what is bone-dust. The powdered skeleton of an animal. How did the animal make the skeleton? By taking in its food the minerals which it required. In truth, dear reader, did you not swallow bone-dust at some period of your life you would now be in a sorry plight. Did you ever hear of a disease called "rickets?" No; well, it is a state of the system when no bone-dust has been made into bones; and in this disease one can virtually put one's feet in one's pockets, for one has no longer solid thigh or leg bones, but flexible, as if they were made of india-rubber. So you see we *must* eat minerals, and the principal sorts we require are the following:—

    Common salt.
    Phosphate of lime.
    Phosphate of magnesia.
    Limestone, or chalk.
    Carbonate of soda (washing soda).

The first has been employed as aliment from time immemorial; we find it mentioned in the Bible as having been most extensively used by the Jews. "If the salt hath lost its savour, wherewith shall it be salted?" Eastern travellers also describe it as being a mark of respect on the part of potentates to present each other with salt. There is no portion of body in which it may not be found. Even among the lower animals the desire for salt is often very powerfully shown; for instance, in some of the wild districts of America there are portions of the soil strongly impregnated with common salt, and it is so well known that in certain seasons these localities are thronged by hundreds of buffaloes and other wild cattle, seeking this compound, that they have been called the "Buffalo licks."

The other mineral ingredients are abundant in vegetable food, as corn, bread, potatoes, cauliflowers, &c. &c. The young of Mammalia* are dependent for subsistence upon the milk of the mother, and as this liquid maintains life and forwards the growth of the being for some time, an inquiry into its composition and properties may aid us in forming a general idea of the kinds of food which such an animal requires. Milk is composed of—

    Water,
    Butter,
    Cheese,
    Sugar,
    Common salt,
    Phosphates and carbonates,

* Animals which have teats, and feed their infant offspring upon milk. From the Latin *mamma*, a teat.

which are the very articles of diet that every one, high or low, consumes from day to day; for though albumen (white-of-egg) be not here present, we may look on cheese as only a very slightly modified form of this principle, and one which is converted into a substance similar to albumen, prior to its absorption into the blood through the vessels of the stomach.

We have now travelled over most of the food ground; but since there are two alimentary compounds in general use which come under no distinct class, and to which we have not yet alluded, I shall treat of them shortly. Tea and coffee—are they nutritious? I think the answer **must** be in the affirmative, at any rate in the way in which we use **them**. But whether tea and coffee are intrinsically nutritive compounds is **a** question. When they have been submitted to chemical **analysis** two peculiar crystalline salts have been obtained, **which are respectively** called Theine and Caffeine. These contain a **very large percentage** of nitrogen, and therefore would be thought **by the Liebig school** of chemists to be of great tissue-repairing value. **I very** much *doubt* they are *ever* concerned in the reconstruction **of the** soft parts. It appears to me that their chief use is as stimulants of an agreeable description, and which are somewhat preservative; that is to say, have the power to prevent waste of the bodily structures. This view is substantiated by the loss of general appetite experienced by that *most respectable* class of society—the aged tea-drinking females.

It will not be out of place here to say two or three words on the subject of soups and gelatine. An endless amount of misconception exists in reference to these materials. It is stated that soups are innutritious because they contain gelatine, which is also useless as a nutritive substance. I do **not** know of any physiologist who has so philosophically and **interestingly** treated on this question as Mr. G. H. Lewes.* I cannot concur in **all** his opinions, but I fully agree with him regarding the illogical manner in which the conclusions of those who would exclude gelatine **from** the class of nutritious food **are** drawn. **Great** excitement was produced some years since in Paris by the announcement that gelatine was highly nutritious, and could easily be **extracted** from all sorts of bones by a process of **continued** boiling. **In all the** public charities boilers and extensive apparatus for the manufacture of gelatine were erected, and **as** the poor flocked in great numbers and were supplied with the soup extracted in this manner, *apparently* ample opportunities were afforded of testing the truth of the doctrine. But the product proved anything but nutritious, and gave rise to diarrhœa and other unpleasant symptoms of imperfect digestion. Mr. Lewes wittily but truly observes:—
"The *savans* heard this with great equanimity. They were not the men to give up a theory at the bidding of vulgar experience. Diarrhœa was doubtless distressing, but science was not implicated in **that**." And so the affair went **on** from bad to worse, till

---

* **Physiology** of Common Life, in 2 vols. By G. H. Lewes. Blackwood & Co. 1859.

the Academy having been assured of the innutritive qualities of gelatine, decided against it. In 1833 two distinguished Frenchmen brought the matter a second time forward, it was again defeated by the Academy, and so it rests. Gelatine is therefore considered by most modern physiologists as useless food. Is there of my readers any one who has been an invalid, and during convalescence has been confined to a gelatinous diet? If so, I would ask him, has he not felt substantial proof of the nutritive power of the delicious broths and jellies with which his kind nurses have provided him? Does he reply in the negative? Then I take the question myself, and say, *I* most assuredly have been rebuilt bodily by a diet consisting of broths and jellies. Besides, the strongest argument against gelatine is, that it *alone* is incapable of sustaining life. We shall see presently how this also applies to every article of food taken separately, and, therefore, by the same line of argument, we should arrive at the painful conclusion that there was no food in the universe.

Chemists tell us that after gelatine has been taken into the system a large quantity of urea is found in the urine, and that this shows the want of nutritious power in the gelatine; but by what abstruse mode of reasoning they have come to this result we are not informed. That gelatine *is* nutritious I think we have fair reason to suppose, while there is not a tittle of real evidence on the other side. We may therefore take it for granted as extremely probable that soups, broths, jellies, and their kin, are of considerable value as articles of diet.

The questions we have up to this time been discussing relate to food, and lead us by an easy gradation to the question, What is *poison?* I know nothing more difficult to define than this expression, and though hosts of wordy and technical significations are to be found in works on forensic medicine, I believe I am not outstepping truth in asserting that none are faultless. Let us say *a poison is some substance, which, when allowed to act of itself, either within or upon the body, tends to produce death.* Thus arsenic is a poison if applied to the skin, and still more so if taken into the stomach. Pounded glass will produce death, if swallowed, by giving rise to inflammation of the intestinal canal; and, strange to say, pure mercury, even if so much as a pound's weight of it be taken, will give rise to no dangerous effects at all. There are queer notions afloat among the public about the retention of mercury within the system. Be it known then, that three weeks after even the largest dose of blue pill has been taken, not a particle of it can be found within the body. To prove this, men who have been poisoned by mercury, but who lingered on for this period, have been regularly boiled down, and carefully examined, yet never has the faintest trace of this metal been discovered. Even common table salt is a poison, and cases are on record of death from the swallowing of this substance. Again, a salt which one man can resist the action of proves fatal to another, thus showing the truth of the adage "One man's meat is another's poison." Strychnine, whose properties everyone in

England is too well acquainted with, is said to be poisonous to every animal except one, and this is a species of bird of the genus Hornbill, which is said to live upon the nux vomica nuts, from which this "deadly poison" is derived.

Apropos of food, it is a curious fact that unless the stomach be nearly filled, the sensation of hunger remains, but if with the aliment there be mixed mineral matter which is entirely indigestible, the feeling of starvation vanishes. Thus Humboldt says of the people of Oronoco that they often mixed earth with their food, and when asked the reason for this, replied that they did it because they knew the stomach should be filled. In the famine which occurred in 1832 on the borders of Lapland, the people of Degernä formed a kind of bread by baking together flour, the bark of trees, and a peculiar clay, which, when analysed, was found to contain nineteen different forms of Diatomaceæ * in the fossil condition. What a little treasure for some of our enthusiastic microscopists the bellies of these poor Laplanders would have been.

No animal can continue to exist if provided with but one and the same article of diet. This has been proved by feeding dogs upon sugar alone, the result of which has been emaciation, loss of appetite and strength, listlessness, the appearance of ulcers in the eyes, and death after from thirty-one to thirty-six days. Magendie fed a dog on white bread and water, and found that he survived but fifty days. Rabbits, guinea pigs, donkeys, fowls, and dogs have been fed upon *one* only of the following sorts of food:—wheat, oats, barley, cabbage, rice, carrots, hard eggs, cheese; and death has invariably supervened in a longer or shorter period. We often talk in this country of keeping prisoners on bread and water diet, but few are aware in how short a time this would produce death. In Denmark it is so well known, however, that a diet of bread and water for four weeks is considered to be synonymous with capital punishment.

Dr. Stark (who, being desirous of ascertaining the effects of restriction to a diet of one kind, began a series of experiments upon himself) discovered, when too late, the truth of the above statement, and perished in his honest endeavours to advance science. Oh, ye self-experimenting physiologists of the nineteenth century, let this be a dreadful warning to you! Beware of unintentionally immolating yourselves at the shrine of biology.†

Having waded through so much on the varieties and actions of food, the reader may very naturally inquire what is the cause of hunger. And here I am at fault, for I know not what it is myself; and furthermore, I am quite certain that those writers who give such diffuse explanations of the mattter, are themselves as completely

* So called from two Greek words, meaning to cut through; it being the character of these lowly and flinty plants to split up into segments of various shapes.

† The science of life; from two Greek words, *bios*, life, and *logos*, a discourse. Unhappily, the progress of charlatanism in this country has created an impression in the public mind that the above word has some connection with mesmerism.

ignorant. All we can say at present is, that hunger is a peculiar sensation conveyed to the mind by the nerves of the stomach, and caused, *possibly*, by the swollen state of the vessels of this organ, owing to the absence of food. But this is really little more than the reader is aware of already, and he must "take the will for the deed." Better to tell the naked truth, than to lie in using long and obscure sentences to conceal our ignorance.

We are in the same condition of difficulty in regard to the terms thirst and satiety. No precise or satisfactory account of the production of these sensations has as yet been given. Several experiments have been undertaken to discover the length of time for which animals can support starvation, but the results have been practically useless. We are justified, however, from the records of shipwrecked mariners and forlorn adventurers, in arriving at the conclusion that a man suffering hunger and thirst cannot live above a week. If deprived of food *alone*, he can exist for a still longer period, and if he be insane, he may continue to live for an extraordinary length of time (?). This general law we may look on as established. *Warm-blooded animals (mammals and birds) tolerate starvation and thirst for a much shorter duration of time than those whose temperature is lower, as fishes, amphibia (Frogs, &c.), and reptiles.*

Ere I conclude this chapter let me offer a few remarks on the much debated question—Is tobacco a poison? In the *Cornhill Magazine* for November, 1862, a most valuable article appeared on this subject. The writer's signature is not attached to it, but evidently the author is one highly competent to handle the question, and I conceive he has handled it in a masterly manner. Turn then, dear reader, to the periodical alluded to, and fancy, if you like, that I am nodding acquiescence in the statements therein contained, as you scan or study them according to pleasure. Tobacco, whether in the form of cigar or not, contains a peculiar principle of a narcotic power, which is called, after the plant itself,* nicotine. In addition to this, when burnt (as in smoking), an oily compound is formed, which is poisonous. These two materials are not necessarily absorbed to an injurious extent by the tissues of the body. Therefore, moderate smoking is not, as some would have it, a most pernicious habit. We are told it expends the saliva; be it so, what of that? Oh, my dear sir, it is one of the most important fluids employed in digestion. *Quære*—Is not the office of the saliva a purely mechanical one, softening the food to render chewing facile, and oiling, as it were, the gullet, so that the mass may slip easily into the stomach? Modern research tends to shew the truth of this. But, sir, tobacco stimulates the nervous system. I say, granted; and a very delightful stimulant it is; whilst, my sage physiologist, I am to suppose that tea, coffee, wine, beer, spirits, *pleasure*, are not

---

* The tobacco plant belongs to the same natural order as the potato, henbane, and bitter-sweet, and is termed in botanical parlance, *Nicotiana tabacum*. Apart from its general use, it is of great value as a medicine,— for example, in cases of rupture.

stimulants also. Eh? Again, dear reader, let me advise you never to let fungus philosophy triumph over common sense; and when your pigmy savant tells you that men are smoking away their brains, point to Germany, and say, Is there no profound thinking there? Is there no mental power in men who, as Thomas Carlyle, Lytton Bulwer, and Thackeray, are known not only to indulge, but to encourage indulgence in that solace of the student—the fragrant weed?

CHAPTER VI.

Digestion—What we require in order to digest Food—All Animals have Stomachs—How the Amœba eats and digests—Gullet, Intestine, and Teeth are Appendages—Glands employed in Digestion—Varieties of Teeth—Structure of a Tooth—Mastication—The Spittle Glands—Structure of the Parotid—Use of the Saliva—Experiments of Bernard—Discovery of Lassaigne—Influence of the Mind upon the Secretion of Saliva—Form of the Stomach—Glands which form the Gastric Juice—Characters of the latter—Pepsin—How Food is dissolved—Fat is not dissolved in the Stomach—Fats and Starch digested in the Small Intestine—Nature and Office of a Villus—Intestinal Spittle Glands—Function of the Intestinal Juice—Structure of the Liver—Use of Bile—The Sweetbread—Nerves which supply the Guts—Length of Intestine dependent upon the Character of Food required—Movements of the Stomach—Absorption of Food—Chyle—Form, Position, and Use of Lacteals—Time required for the Digestion of various forms of Food—Quantity of Gland Juice formed in Twenty-four Hours—Amount of Food required per Diem—Indigestion—Dyspepsia—Cleanliness.

ALL of us know, in a general way, what is understood by the word digestion; but perhaps if some were asked what was necessary to ensure digestion, they could not give a very satisfactory reply. There are three things requisite: 1st, food; of this we have spoken already. 2nd, some vessel or sac to hold the food; and 3rd, certain liquids, to dissolve and act upon the latter so that it may be taken up into the system by the blood-vessels. The second factor—that of a cavity to contain the alimentary substances whilst they are being fitted for absorption—is the stomach. It is the chemical laboratory of the body, without which, as very well shown in the Roman's fable of "The Belly and the Members," life could not be carried on. Here, *en parenthèse*, I may observe what may surprise a few: a stomach, or some approach to it, is the only character by which the lowest plants and animals can be contra-distinguished. Most animals have a stomach, or gullet of some kind; but even those which possess no such commodity extemporize one when they require it. *Plants neither possess nor extemporize a digestive sac.*

Here is one of the most degraded of animal forms—the little "Amœba"—a creature to be found in pools of fresh water in the summer time, but not so easily discovered as some of our naturalists would lead us to imagine. Watch him closely! He is but an irregularly-shaped mass of jelly, and very minute; moreover, he is quite transparent, and is not (as our quack advertisements have it) "troubled with a liver," nor indeed with any organ at all. He is really, as a lady once remarked to me of a snail—"all squash;" and yet, as I said before, watch him! Lo! a minute animalcule has just brushed past him. Ah, luckless animalcule, not *past!* for the Amœba has thrown out a long whip-like string of jelly, in which thou art entangled. Struggles are unavailing. The relentless monster has seized his quarry; and see! already he is throwing out other arms, hydra fashion—now two, now four. In a moment a dozen hungry arms have closed around thee. Stay! "What will he do with it?" Art thou to be kept "in durance vile," hapless infusorian? 'Tis true thy cruel tyrant hath got no Bastile in which to entomb thee; a far worse fate is being prepared. The Amœba is gradually pushing his prey into the substance of his body. This has subsequently closed over it; and what do we behold? A transparent sphere enclosing the unfortunate animalcule, who is now subserving the comfort of the oppressor by undergoing rapid digestion. After some time, when all the nutritious materials have been abstracted, the remnants are quite *unconcernedly* forced out through some portion of the gelatinous film.

Fancy a little girl making her first dumpling, by placing a piece of apple and sugar upon a layer of dough, and then tucking these in so as to form a roundish ball of paste, and you will understand very clearly the *nonchalance* exhibited by our Amœba friend in getting through his meal. A stomach, then, is the simplest form of digestive cavity, and the most essential; all the other organs of the food system are but superadded. Thus, as we advance in the animal scale, we find the stomach placed in the lower part of the body. Therefore it is necessary that there be a channel through which the food may reach it; this, then, is called a gullet, and its outer opening the mouth. Next there must be provided a canal, by which the indigestible *débris* are carried away; this is the intestine. Again, the food of some animals is of a dense character, and must be firmly divided before entering the stomach. To overcome this difficulty teeth are presented; and so on, till we find in man the machinery existing in its most perfect and complex form. In the human digestive apparatus we have, in addition to the above components, certain *glands.* Glands are organs composed of the different tissues arranged in a special manner, whose office it is to separate, or form from the blood, fluids which are some of them of value, others effete. The principal glandular structures are those of the mouth, which secrete the spittle, and are

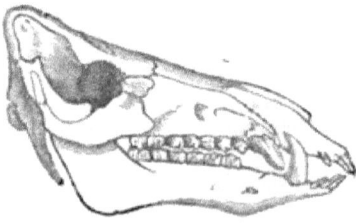

Fig. 10A.—Skull of the Boar, showing the three classes of teeth.

STRUCTURE OF TEETH.

called salivary; the liver and sweetbread, which pour their secretions into the top of the intestines; the sacules, which form the gastric juice, and glands called after Brunner, which are found in the smaller gut. The teeth of an adult are thirty-two in number—sixteen in each jaw, and are of three kinds—incisor or gnawing, canine or chopping, and molar or grinding (fig. 10B). These varieties characterize dis-

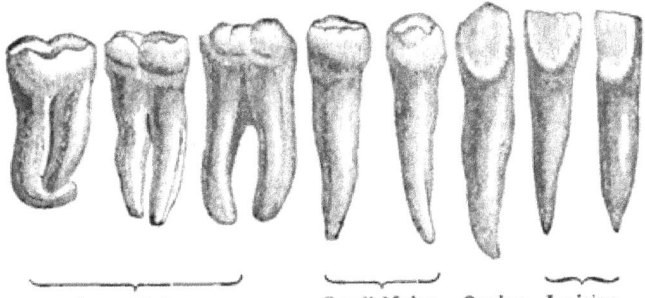

Fig. 10B.—Human Teeth.

tinct orders of Mammalia. The incisors are found in animals of a gnawing propensity, as rats, rabbits, beavers, &c., which are termed Rodents. The second form of teeth is especially indicative of a flesh-eating habit, being found most perfectly developed in carnivorous beasts, as the dog (*canis*), whence the name canine. The molars point to an herbivorous diet, and are the principal teeth in such grass-eating creatures as sheep, cows, and goats. To the reflective mind, the mixture of these classes of dental organs in man is very significant, for it shows that the "lords of the creation" were not intended to subsist entirely upon either animal or vegetable diet; and it is one of the strongest arguments against vegetarianists. A tooth is composed of three structures, differing in hardness (fig. 11). The outer—a very thin layer, and the most dense animal substance known—is named enamel; it forms the crown of the tooth, and is to it what the steel edge is to the iron weapon. Next comes the great bulk of the tooth—a structure similar to ivory, and called dentine. It is hollow within, having a cavity that, in the living tooth, is filled with a mass of fat and nerve. It is prolonged upward as the fang or fangs which fasten the tooth in the jaw, just as if it were a nail driven into a board; and these are covered outside by the cement—a bony compound, which often is continuous with the osseous tissue of the jaw, thus enhancing the agony which the patient who is having a tooth extracted must undergo. You may be asto-

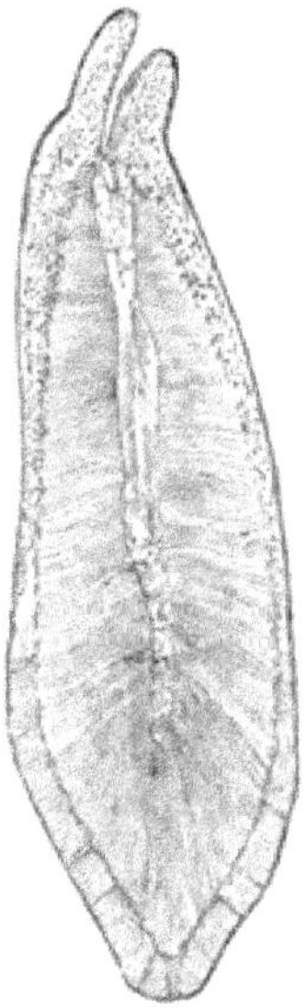

Fig. 11.—Section of Tooth.

nished to learn that teeth and hair are near relations, being "brought up" in the same manner, and having a structure in many ways similar. Indeed feathers and porcupine quills come under the same brotherhood also.

There being three kinds of teeth in the jaws of man, it follows that as many distinct forms of motion must take place during the process of mastication. The upper jaw being fixed, it is the lower upon which the labour of dividing the food chiefly falls. This may be moved backwards or forwards, up or down, and from right to left, by muscles for the purpose, as shown in the adjacent diagram (fig. 12). The tongue plays an important part in the chewing of food; by reason of the muscles of which it is composed it tilts the food-mass from side to side, so that the large particles are retained beneath the teeth until reduced to a sufficiently fine condition.

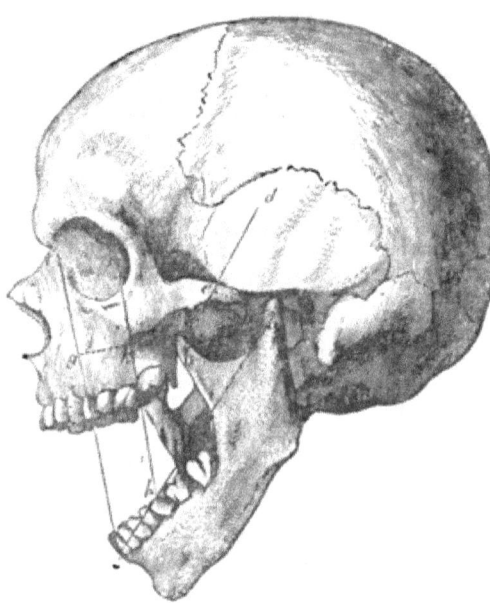

Fig. 12.—Human Skull, with the lower jaw (*a a*) depressed; *g k*, the upper jaw; *d*, surface of the skull, to which one of the muscles (temporal) is attached; *b*, projection of lower jaw, in which the same muscle is inserted. The vertical and oblique lines represent the directions in which the jaw can be moved.

While the process of mastication is going on, six glands are actively engaged throwing out saliva into the mouth; and, ere we consider the use of the spittle secretion and its composition, we must glance at one of the organs which develope it. The six glands are so very like each other that one description must answer for all. Tracing one from the mouth, it appears as a membranous canal; this soon divides and subdivides, like the branches of a tree, till there are produced at last exceedingly minute tubules. If, now, we examine these under the microscope, we see them expanding at their extremities into little bladders or sacs, which give the whole gland the appearance of a cluster of grapes in miniature (*vide* fig. 13). Between these sacs there are hosts of microscopic blood-vessels, both arteries and veins, and finally all the little nooks and crannies are filled up with fat. The secretion is derived from the blood circulating in the delicate vessels with which the little sacs are surrounded; in fact, it is but a *kind* of percolation of the more liquid and saline portions of this fluid, through the coats of the vessels and sacs, into the cavities of the latter, from which it is then conveyed to the mouth by the series of conduits I have described. If you have borne in mind the remarks I have made in another chapter on the nature of *osmose*, you will then, by the assistance of the diagram (fig. 14), be able readily to comprehend the process.

The saliva is a somewhat viscid liquid, and contains, besides water and saline materials, a quantity of a leavening substance called ptyaline. It is usually alkaline, but in some cases has been found acid. Its uses are :—

1. To soften the food, so as to admit of its being more easily chewed.
2. To mix up air with the food.
3. To moisten the throat and gullet, so that the food may glide freely to the latter.

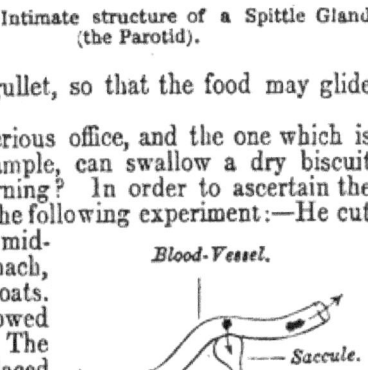

Fig. 13.—Intimate structure of a Spittle Gland (the Parotid).

The third seems to be its most serious office, and the one which is most evident also. Who, for example, can swallow a dry biscuit with ease, on first rising in the morning? In order to ascertain the value of saliva, M. Bernard tried the following experiment:—He cut across the gullet of a horse about midway between the mouth and stomach, and then gave him a pound of oats. This the beast chewed and swallowed in the course of *nine* minutes. The food was received into a vessel placed under the cut portion of the gullet, and when examined was found properly masticated and saturated with saliva.

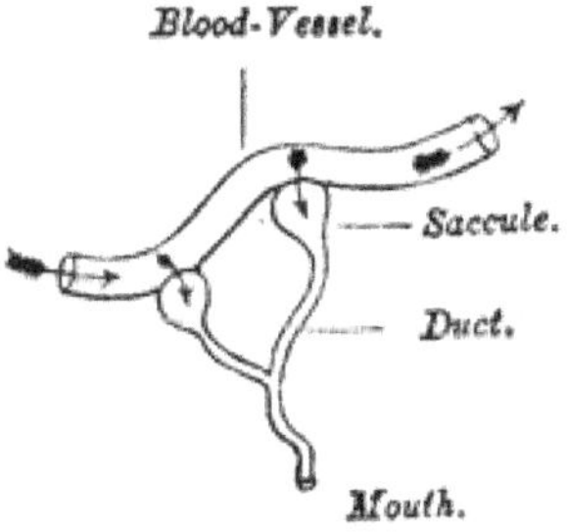

Next, he caused the spittle to flow out of the mouth through an incision, and administered another pound of oats. On this occasion but three-quarters of the food were eaten, and no less than twenty-five minutes were expended in swallowing them. The mass, which as before was received into a vessel, was this time found dry and brittle, and entirely devoid of saliva.

Fig. 14.

It appears that the quantity of saliva poured out depends upon the character of the food; if this be hard and dry, there is a large supply of spittle, and *vice versâ*. M. Lassaigne proved this by an experiment rather like the one I have just mentioned. A cow was supplied with *weighed* quantities of four kinds of food, and an incision having been previously made in the gullet, these were, after swallowing, collected and re-weighed, when, of course, the excess indicated the amount of saliva employed in each instance. Thus:—

For 100 parts of hay, there were absorbed 400 parts of saliva.
,, ,, barleymeal, ,, ,, 186 ,, ,,
,, ,, oats, ,, ,, 113 ,, ,,
,, ,, green stalks and leaves, 49 ,, ,,

The above researches afford a very fair conclusion as to the functions which I have attributed to the saliva; but that it fulfils any more useful office, I think we are at liberty to consider very questionable.

The flow of spittle fluid is influenced to a great extent by mental sensations. The appearance of delicious viands is often sufficient to produce an increased quantity of the secretion, owing to the effect upon the mind; of which the proverbial expression, "making the mouth water," is a very good illustration. Saliva possesses the power of converting starch into sugar. This action takes place when farinaceous food has been retained in the mouth for some time; and will occur out of the body, in a glass vessel for example. Physiologists have been led from this circumstance to suppose that it performs the same part in the stomach; but experiments have done away with this notion by showing that the juices of the stomach completely prevent the alteration.

All this time I have been keeping you with a bit in your mouth, which may now be swallowed if you please; but in doing so you must not forget that the road to the stomach is the gullet, and if your food parcel is sent "the wrong way," the excise of the wind-pipe will very soon inform you of its dereliction of *duty*. Now comes the gullet, and with it the stomach and guts, or bowels, as they are more frequently designated.

We may look on the entire alimentary "line" as a fleshy tunnel or tube, blown out at the stomach into a sac; in the same manner as the London and North-Western Railway, lying between Chester and Euston Square, expands into the magnificent station in Birmingham. The whole digestive canal, from mouth to opposite extreme, is made up of three layers; a soft, sinewy one outside, a fleshy one in the middle, and a beautifully piled velvet lining within.

The stomach lies at the top of the belly, and in front; and has been often compared to a piper's bag in form (fig. 15). It is joined to the gullet at its left and upper border (*a*), and merges in the small intestine on its right side (*vide* diagram). It is by the stomach that the gastric (*gaster*, the stomach) juice is secreted, which is of so much import in digestion, and which we hear dyspeptic sufferers complain of so often. How is this stomach fluid formed? By little pouches lying in the velvety coat, which are generally shut up while we are fasting, but as soon as food is introduced throw out their peculiar secretion. The whole inner surface of the stomach is lined with these, which have the appearance seen below when examined microscopically (figs. 16, 17).

The gastric juice is of an acid character; but whether this sourness is due to the presence of vinegar, spirits of salts, acid of milk, or acid of rancid butter, is yet a question; indeed, it is quite possible it may be owing to all of these. Its most important and essential constituent is a substance termed pepsin, which acts upon all fleshy foods as a ferment, reducing them to the state of white of egg.

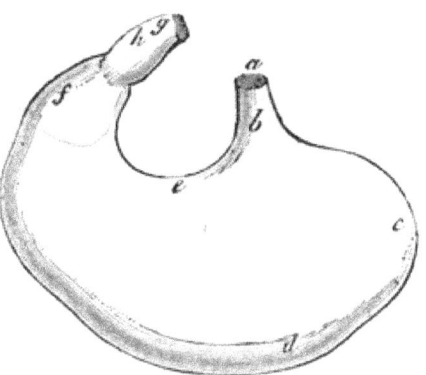

Fig. 15.—Diagram of Human Stomach— $a\ b$, gullet end; $h\ g$, intestinal end; c, left border; f, right border; d, greater curve; e, lesser curve.

It is not improbable that this organic compound, pepsin, may act upon the common salt, so as to cause its decomposition, and give rise to acid. Thus:—

In the stomach during digestion are
1. Flesh ——————————————— White of egg.
2. Common salt { Sodium
                 Chlorine } Acid Soda
3. Water .. .. { Hydrogen
                 Oxygen

I cannot here go into further details, and must only hope that the scheme above may help to supply the want of further explanation. Acids do not dissolve oils, therefore, as we might expect, fats undergo no solution in the stomach. Flesh, and all liquids except those of a purely oleaginous character, the albuminous parts of plants and bread, and saline materials are dissolved in this locality. Fat and starchy matters pass into the intestines before they assume the soluble form. The intestines, whilst partaking of the structural features of the gullet and stomach, have many peculiarities. All through they are lined with the velvet coating and small pouches, which in the stomach secreted gastric juice, but which here manufacture a viscid liquid named mucus. This lubricates the food-mass as it travels along, and so prevents injury to the membrane.

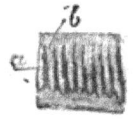

Fig. 16.—Gastric glands from a man's stomach— a, glands; b, velvety membrane.

The velvet surface is raised out in little conical projections, termed *villi* (fig. 18 B), and these, as we shall hereafter see, are the labourers employed to extract the nutritious materials from the crude combinations of useful and effete. Each *villus*, seen under the microscope, resembles a little teat, covered with a velvet layer outside, and having within, embedded in loose tissue, a delicate network of blood-vessels, and a small white tube called a lacteal, — of which more anon.

Again, buried in the walls of the lesser intestine is a series of glands rather like those which secrete the saliva, and of the form represented below (fig. 19). The fluid created by these organs is similar in properties to that found in the mouth —of which we have seen enough already.

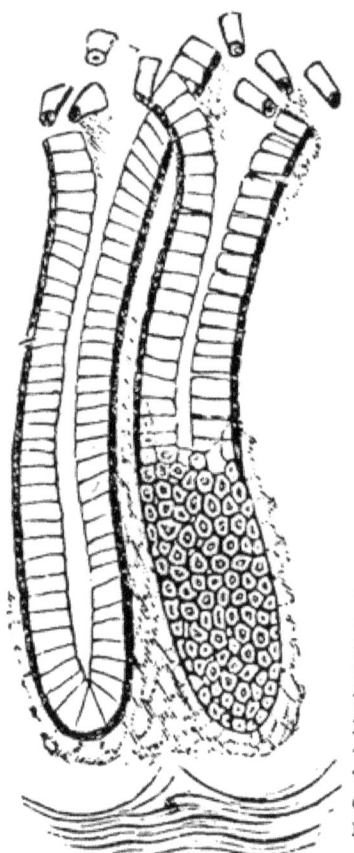

Fig. 17.—Gastric glands from the stomach of a Pig, highly magnified.

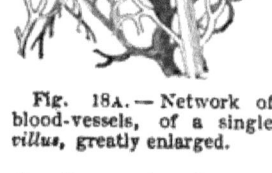

Fig. 18A.— Network of blood-vessels, of a single *villus*, greatly enlarged.

In addition to those already mentioned, there are found all through the intestines, small oval white solid bodies, surrounded by a circle of minute apertures, but of their functions we really know nothing yet; they are collectively called the glands of Peyer. All these lesser intestinal glands contribute to form a fluid known by the name of "intestinal juice." That it also operates on flesh has

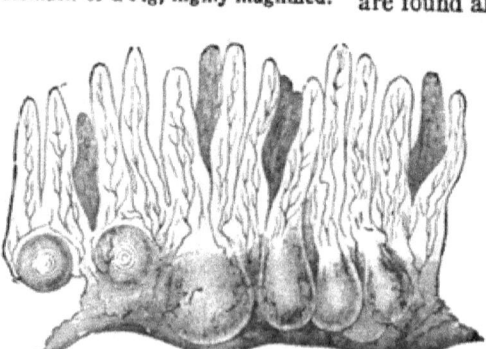

Fig. 18B.—Clusters of *villi*, magnified.

STRUCTURE OF THE LIVER.

been demonstrated by two German chemists in the following manner. They made an opening into the belly of a dog, and drew out a portion of the intestines, which they then tied near the stomach, to prevent the gastric juice coming down; next, they opened into the intestine, and placed in it two or three muslin bags containing weighed quantities of meat, and replacing it in the belly, allowed the animal to live for about twelve hours. It was then killed, and when the bags of flesh were extracted, and re-weighed, it was found that the contents had been altered, and the weight diminished.

Up to the present we have omitted considering the two largest organs in connection with the digestive system; they are the liver and sweetbread. The liver is a very large brownish gland, which lies just beneath the fleshy partition (diaphragm) separating chest from belly (vide fig. 21). It is composed of a vast number of tubes, which are derived from a central one, being given off from this latter in an arborescent manner. The main tube, which is more commonly known as the bile duct, opens into the small intestine below the stomach, and the ends of the minutest hollow branchlets are closed or blind.

Fig. 19.—Spittle glands of the intestine, enlarged.

Besides this tubular apparatus there is a mass of soft solid material, which gives the shape to the organ, and with the veins and arteries which are present also fills up the spaces between the different tubes. If a thin slice of the solid matter be placed in the field of the microscope, it presents this appearance (fig. 20). It would take too long to enter fully into the structure of this organ, which is, however, most interesting to the microscopic anatomist. The fluid formed in the liver is called bile, and filters from the blood into the blind pouch-like ends of the tube branches, from which it flows into the larger ones, and from these again into the central duct.

Now, there is attached to the liver, and opening into this central duct, a small bladder, which serves as a reservoir to contain the bile as it is gradually distilled; for we find that, when digestion is not going on, the bile does not flow into the intestine, but, after having passed into the bile duct, is sent backwards into this reservoir, the gall-bladder (fig. 21).

Some physiologists have thought that the liver, besides secreting bile, altered the blood passing through it by giving it *sugar*. The notion has recently been completely upset, so that though this organ may fulfil other offices besides the formation of bile, the latter is all we can maintain to take place in it.

The ancients believed that the liver exercised a powerful influence over the mind; and our common word melancholy is derived on this supposition from two Greek words which signify black bile. There is no slight reason to suppose that the state of the mind does, in some measure, depend upon the condition of the liver; for, if this organ be disordered even to a small extent, the effect is very soon observable in the general system.

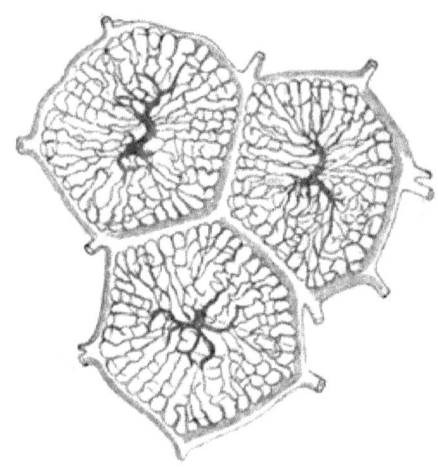

Fig. 20.—Structure of Liver.

The physician will tell you, "Your liver is out of order, no bile is formed, and you can't in consequence digest your food;" but push him a step further, ask him why, and he is floored at once. If we take a portion of boiled meat—fat and flesh—and place it in a vessel with some bile, keeping up the temperature at the height usual in the body, about 98° Fahrenheit, we shall find that no change takes place. Therefore, it is evident that it has no *direct* action on the food. But if, on the other hand, we make an opening into the gall-bladder, so that the bile passes out through the wound, and not into the intestine, the digestion goes on very imperfectly, there is constipation of the bowels, emaciation, and death. This experiment has been performed on dogs with the result I have given; but strange to tell, if the animals were permitted to lick the wound, and so swallow the bile, digestion went on nearly as in the natural state; hence it is quite clear that the bile is of *service*.

It is also injurious; for, if it is not secreted, it accumulates in the system, staining the skin and eyes, and terminating in the extinction of life. This is what we see in jaundice.

COMPOSITION AND USE OF BILE.

Comparative anatomy is no mean teacher in a case like this. It shows us that in the lower creatures the liver conducts the bile into the stomach; whereas, if it were only sewage or refuse matter, it would be poured into the lower division of the gut, and not into the digestive cavity, where it might be absorbed, and if effete

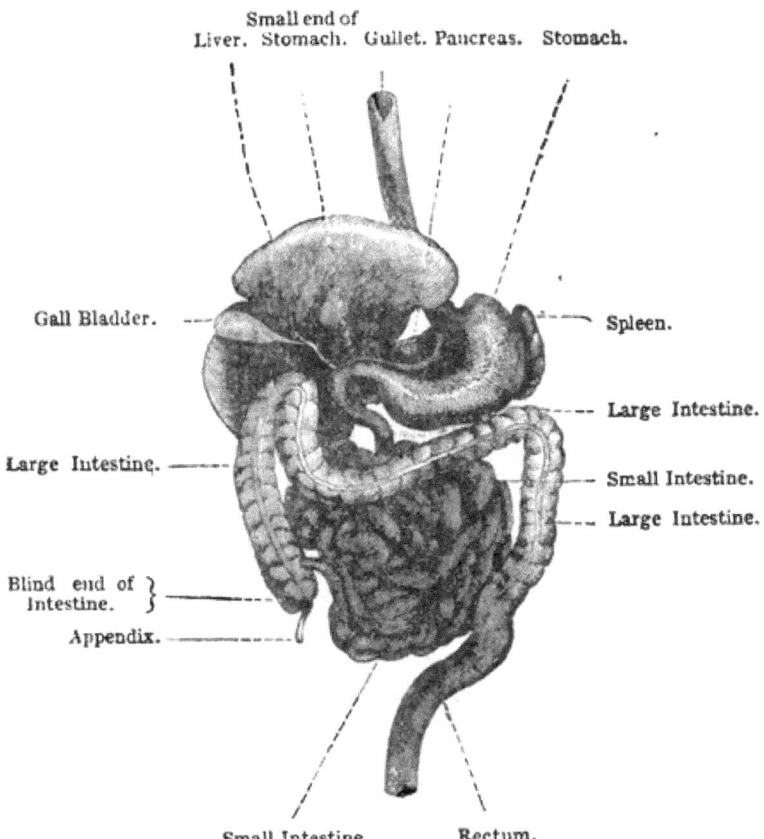

Fig. 21.—Digestive Apparatus in Man.

might act poisonously. All, then, that we can say of the bile is, that it certainly is of use, but that we have not yet discovered *in what way*. It is composed of—

Water............ 880
Bile matter 90
Colouring matter } 15   in 1000 parts.
Fats .... }
Salts 15

1000

That the bile does not pass away as such with the excrement is certain, for the quantity of sulphur which the latter contains does not amount to one-eighteenth of what is poured into the intestine with the liver fluid. What becomes of it?—that is a mystery to physiologists.

Next we come to the consideration of the sweetbread or pancreas (from two Greek words, meaning all flesh). This is a long flat gland, lying on the back of the stomach, and whose structure we need not here discuss, as it is similar to that of the spittle glands already described. The channel which conveys away the sweetbread's secretion, opens into the small gut below the stomach, and exactly opposite the opening of the bile duct.

The liquid formed by this organ is transparent, viscid, and alkaline, strongly resembling the saliva in these particulars. Its action is confined to the fatty and starchy matters of the food. It converts the starch into sugar, and by causing the oily substances to be reduced to a state of very fine division, renders them soluble. How is this demonstrated? Remove from the pancreas of a dog a quantity of the secretion, place it in a glass vessel with some oil and starch, maintain a temperature of 100 degrees Fahrenheit, and after some hours you will find the starch turned into sugar, and the oil changed to a whitish liquid, called an *emulsion*.

But this is not all. M. Bernard has proved *conclusively* that the above is the office of the pancreatic fluid. The rabbit was the animal he selected for his experiment, because in it the canal of the sweetbread opens into the gut much lower down than that of the liver (fig. 22). He tied the intestine at *a*, having previously introduced a quantity of fatty and starchy food, and he also tied it at *b*, thus leaving a sac of gut, *c*, between the two knots, into which sac the secretion of the sweetbread was carried. After a certain number of hours the animal was killed, and the food was examined, when it was found that the fat had been in some measure dissolved and absorbed, and the starch was to a great extent converted into sugar.

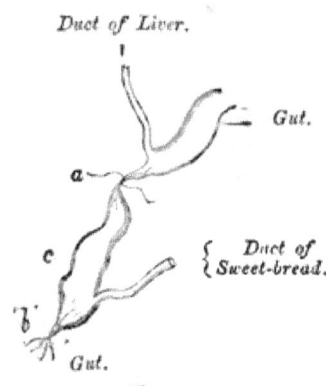

Fig. 22.

The pancreatic juice consists of

Water. 900
A peculiar substance,
 called Pancreatine 90 } in 1000 parts.
Mineral Salts . . . 10
———
1000

Before we come to the subject of digestion as a whole, let me observe that the intestinal canal is not dependent on the general nervous system for its supply of nervous power. A German anatomist has lately discovered a most complete series of nerve centres, situate between the layers of the guts along their entire length; and though, doubtless, this luckless individual will receive many a blow ere his views are recognized, I have great confidence in the truthfulness of his observations. The intestines do not lie loosely in the belly, but are gathered up by a membrane attached to them and the backbone. (Fig. 23.)

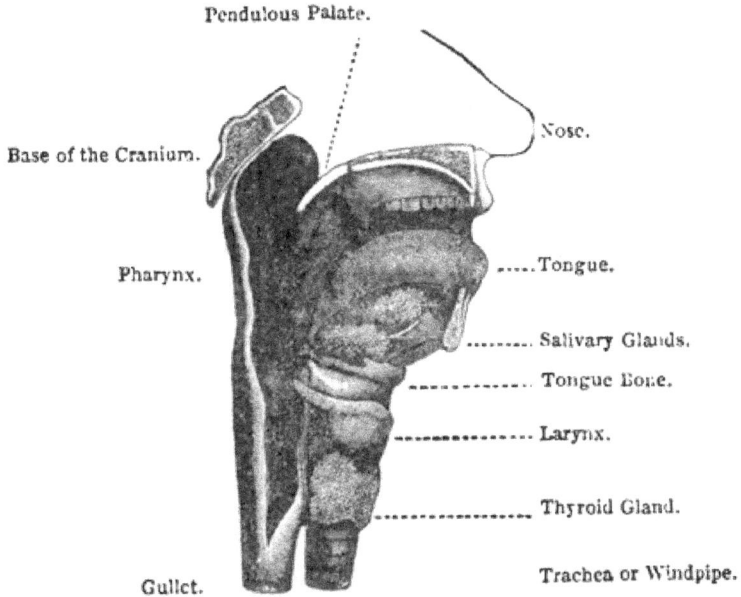

Fig. 23.—Vertical Section of the Throat and Mouth.

The intestinal canal is exceedingly long in herbivorous animals; being in the sheep twenty-eight times the length of the body. In flesh-eaters, on the contrary, it is extremely short; thus, in certain blood-sucking bats it is little more than the length of the creature itself. The reason of this difference is obvious; in carnivora the food is easily digested, and the process takes place in the stomach; whilst in those animals feeding on plants, the main part of the digestive process is gone through in the intestine.

In man we find the alimentary tract intermediate in dimensions, a circumstance which is clearly indicative of the kind of food required by the human race, [viz., a mixed one,] and is another fact to quash vegetarianism. Man's intestinal canal is about five times the length of his body, which the reader will recollect in association with Abernethy's remark to the lady, " You'll please, ma'am, to remember that you have twelve yards of guts."

Having now gone over the individual wheels and levers of the complex digestive mechanism, we are in a better position to study its movements collectively. How shall we begin? Shall we observe the bolus during its transmission through the *isthmi faucium*, noticing the approximation of the palato-glossi, the elevation of the laryngeal apparatus, depression of the epiglottis, and tensed condition of the *velum pendulum palati?* No, "no more of this, Hal, an' thou lovest me."

Reader, step down with me to yon tavern, and we shall have a chop and roast potatoes and glass of "bitter" together. What! you really will consent to so plebeian an undertaking? Good! Now, whilst you are eating and resting—as you con the last *Punch*, I shall explain to you, by the help of my physiological "Bradshaw," the route of that same mutton chop and those roast potatoes, shewing you how far they travel together, and by what lines they, after separation, arrive at their respective destinations.

Firstly, the mixture of flesh, fat, potatoes, and beer having been well chewed, and mingled with saliva, is turned back by your tongue into your throat (fig. 23), where it is caught by the muscle of the gullet, and gradually forced down into the stomach. (There, that piece of potatoe was too large, and you feel it all the way down, as if it was a bit of stone!) Secondly, the mass has reached the digestive bag; now the entrance and exit openings of this sac close on it, so it is, *pro tempore*, imprisoned. The stomach immediately begins contracting, and the food, together with the gastric juice, is turned round and round from left to right. Whilst this is going on, the flesh, and the albumen of the potatoes are being made quite fluid, and as quick as they are made so, pass into the veins of the stomach, penetrating the wall of these by *endosmose* (*vide* chap. I.).* The beer has passed away in a similar manner, till

* The velocity with which the blood travels along the vessels of the stomach has much to do with the absorption of liquids. This may be proved experimentally in the following manner:—Let a piece of intestine, *a*, be attached at each end, as shown in the diagram, and placed in a vessel containing a strong solution of sugar, *d*; if, then, water be caused to flow from the funnel *b*, through the intestine and tube, *c*, into

there is nothing left but the starch of the potatoes and the fat. This latter is contained in little bladders of fleshy material, which are now

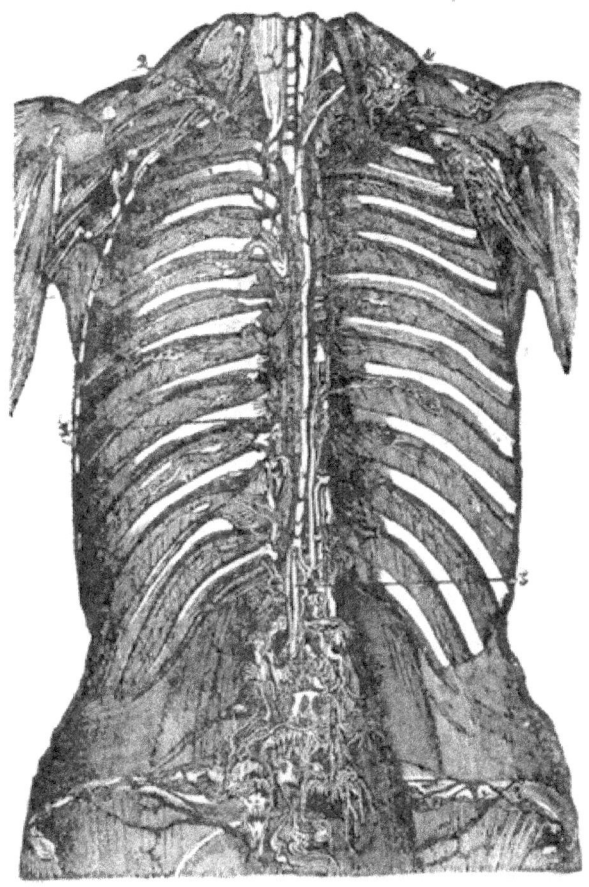

Fig. 24.—The Chest opened, to show the Backbone, and posterior half of the Ribs, with the great Lacteal Vessels. 1. Main trunk of the Lacteal Tree. 2. Lacteal vessels of the body generally. 3. Receptacle of the Chyle. 4. Junction of the Veins of the Shoulder and Neck, and connection of the main Trunk with the Vein of the Left Shoulder.

dissolved by the gastric juice, and follow the beer and potato albumen. Finally, just as the stomach has ceased revolving, there remain in it but oil, starch, and a few bits of tendon; these it has

another vessel, e, it will be found,—1st. That a portion of the saccharine solution has passed into e, and 2nd. That the greater the rapidity of the current, $a\ c$, the greater has been the amount of sugar absorbed.

no jurisdiction over, so the portcullis is opened, and they are transported, after three hours' detention, to the intestine. Here they are met by the pancreatic juice in all its alkalinity, and become speedily transformed, the starch into sugar, the oil into emulsion, and both constitute a white, milky fluid, called *chyle*, which is quite soluble, and is only awaiting my description of the absorbing organs prior to being received into the system.

I have spoken before of tubes called lacteals, which are found in the villi of the intestines. These lacteals are but the terminal

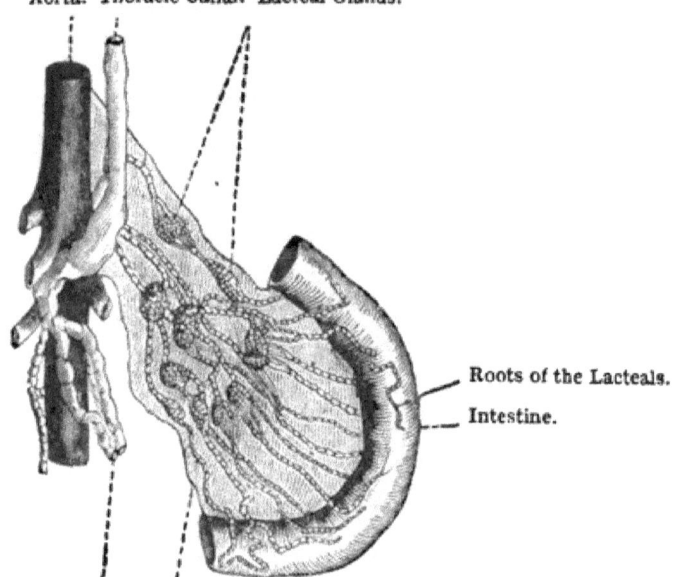

Aorta. Thoracic Canal. Lacteal Glands.

Roots of the Lacteals.
Intestine.

Lacteal Vessels. Membrane sustaining Intestine.

Fig. 25.

branches of a great tree, whose root is at the junction of the two principal veins at the left side of the neck (fig. 24). The stem of this tree lies close to the backbone, as it passes from the neck to the belly, and does not assume the arborescent form till it has reached the membrane which sustains the intestines (fig. 25). It then divides and subdivides, the divisions having knots on them called mesenteric glands, and finally, the last branchlets end each in a little villus, which may be regarded as the leaf (fig. 26). As the foliar parts of a plant abstract certain substances from the surrounding air, so do the lacteal leaflets—the villi—remove from the intestine the nutritious chyle which it contains.

These little villi suck up as it were the fluid sugar and fat; but how these penetrate them is still unsolved. One set of observers says, the lacteals have open ends, and so the chyle passes into them; another, that it enters in obedience to the law of *endosmose*. I confess I incline to the latter supposition, but whichever view be correct, one thing is certain, it does gain an entrance; for, if a dog be killed a short time after a meal of bread and butter, and the belly opened, the chyle by reason of its whiteness can be seen traversing the branchlet lacteals towards the main stem, from which it is afterwards poured into the veins above-mentioned.

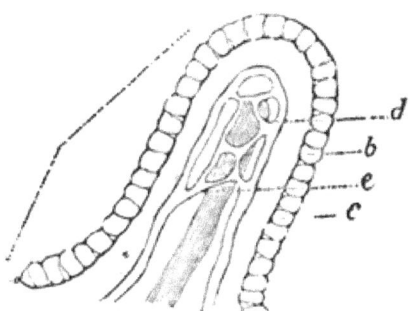

Fig. 26.—Diagram explaining the structure of a Villus; *a*, outer covering of cells; *b*, a single cell; *c*, delicate membrane on which the cells rest; *d*, Blood-vessel; *e*, end of a Lacteal.

If in the confusion consequent upon the discharge of the fatty and starchy matters from the stomach, any particles of undissolved flesh happen to have made their escape, they are placed under the protection of the juices of Brunner's glands (fig. 19). These very soon provide them with the necessary passports to the veins of the intestine; and eventually the refuse substances, bits of tendon, &c., are expelled. Of course, as I have heretofore stated, the bile is also poured into the gut during digestion, but as we do not understand its actions, I omitted it, to avoid confusion.

The period required for the digestion of various forms of meat has been estimated by an American physician, and is as follows:—

Kind of food.	hours.	min.	Kind of food.	hours.	min.
Pig's feet	1	0	Roast beef	3	0
Tripe	1	0	Roast mutton	3	15
Trout, broiled	1	30	Broiled veal	4	0
Venison steak	1	35	Boiled salt beef	4	15
Milk	2	0	Roast pork	5	15
Roast turkey	2	30			

It is not a little surprising to note the quantities of the different digestive fluids poured into the intestine during the twenty-four hours. In weight, the sum of these vastly exceeds that of the excrementitious matters. The following is the amount of each secretion formed in the adult in the course of the day, according to the best authorities:—

	lb.	oz.
Saliva	3	8
Gastric juice	14	0
Bile	3	8
Pancreatic juice	0	6
Intestinal do.	0	7
	21	13

The total is almost astounding, when we consider the small amount of food which is required by a full-grown man, accustomed to outdoor exercise; the entire amount of solid food necessary during twenty-four hours being about two and a half **pounds**. Thus—

	lb.	oz.
Meat	0	16
Bread	0	19
Butter	0	4
	2	7

Of course the actual weight of nutritious aliment demanded will be dependent on the state of the constitution, habits of the individual, and various other circumstances sufficiently obvious. Hosts of people now-a-days complain of their stomachs. You hear that fish disagrees with one, vegetables with another, cheese with a third, and so on, and the term dyspepsia (a compound of two Greek words, signifying difficult digestion) has been applied to this condition.

If the good folk who so very patiently submit to having their stomachs electroplated by the drugs of quack physicians, and allow themselves to be stewed down in those monster human cooking apparatuses, Turkish baths, would only adopt the measures dictated by common sense, I promise them they should know much less of dyspepsia than **they** do at present. For example, my good friends, do not, as the great Scotch surgeon said to the Yankee, "bolt your food like an alligator," and do not overload your stomachs. Especially observe this practice: always, if possible, *rest for half an hour after each meal*. Take a good sponge-bath every morning; most animals wash themselves, but man seldom does so. I should like to know how many men and women clean themselves every day in the week.

The skin, as I shall show you when I come to it, is a great gland of as much importance as the liver or kidneys; and yet how much neglected! I am firmly convinced that, did men regularly wash their bodies on rising every morning, the number of dyspeptic cases would be reduced to one-half. And dyspepsia is a demon, which is at the bottom of many a crime we give other causes credit for. It unfits a man for physical exertion, as well as mental, and by unnerving him, not **un**frequently is the fiend which urges him on to suicide. Again, then, let me repeat, clean yourselves. "Cleanliness and godliness" have ever been espoused. Too much attention cannot be devoted to the bowels, the neglect of which plunges the careless man into fevers, indigestion, kidney disease, and piles — **ay**, piles — those Harpies of middle age! I can add nothing in **force to** the anecdote of Abernethy regarding this subject. "A mother, sending her son **for** the first time to London, remarked, 'My boy, **in** the first place, **have** the fear of God always before your eyes, and in the second, *be sure and keep your bowels open.*'"

CHAPTER VII.

The blood. Its color, microscopic characters and composition—Use of fibrine—Physiology and medico-legal inquiries—Cause of the color of the blood—How does venous blood differ from arterial?—Views of Liebig—Respective actions of carbonic acid and oxygen upon the blood—Blood corpuscles die at the rate of 20,000,000 per second!—Capillaries—Difference between arteries and veins—The circulation of the blood—Structure of the heart—Auricles, ventricles, and valves—Force-pump action of the heart—Other forces concerned in the circulation—Office of the valves in veins—The pulse—Sounds of the heart—The latter a key to the state of the circulatory organ—Action of the mind upon the pulse—Average number of beats in the infant, adult, and aged—Velocity of the blood—Rhythm of the heart—Simile of Harvey's—M. Groux—Dublin experiments—Use of a knowledge of physiology to the physician—Colonel Townsend—Proof of the circulation—Barber Surgeons and Samuel Pepys—Literature of the subject—Did Harvey *discover* the circulation?

WE have seen how the crude materials are converted into albuminous, oily, and starchy substances, required for the manufacture of the tissues and maintenance of the bodily heat. We have yet to consider in what form these substances are best suited to rebuild the tissues of the frame, and also the mode in which they are carried to the distant regions of the animal economy. Thus, in the first place, we shall enter upon the study of the blood itself, and in the second upon the consideration of the way in which the blood is conveyed to the tissues and the power employed in its conveyance—in one word, the *circulation*.

The blood you will tell me is a red liquid? You are wrong; it is perfectly colourless. It is found in every portion of the body, passing through *billions* of little canals so close to one another that the naked eye cannot perceive any spaces between them. Let us collect some of the blood which has just flown from a wound, and place it in a tumbler. After some hours, we see it consists of a red firm jelly of the shape of the vessel, lying at the bottom; and a clear liquid like white of egg floating above it; the first is the *clot*, the second the *serum*.

Now if we still further examine this clot, by taking a very thin slice, washing it well in water, and placing it under the microscope, we perceive that it is formed of a most complicated net-work of white fibres [collectively termed fibrine], which entangle in their meshes several little bi-concave scarlet disks. These latter are the blood corpuscles, and to them is due the scarlet colour of the blood. They are very small and quite imperceptible to the unassisted eye; measuring only about the $\frac{1}{3000}$th of an inch across. These corpuscles are very indestructible, and may be extracted from the blood stains on weapons, upon which the fluid has remained for many years; on which account, as the reader must be aware, the microscopist is able to give material assistance to the lawyer in many criminal investigations.

The supernatant liquor or serum is composed of water, white of egg, and salts, with a trace of fat. That it contains white of egg is evident from the fact, that on boiling it becomes of an opaque white and more or less solid. The fibrine is fluid *only* in the body, for, as soon as the blood is drawn from the wound, it clots, or, as is more usually said, *coagulates*, forming the reticulation described above. What a grand provision is this power of clotting in blood! It is by virtue of this tendency that small wounds close of themselves; a natural sticking-plaster being developed by the fibrine, which plugs up the mouths of the incised arteries. If it were not for the fibrine, the slightest cut would prove fatal; not ligatures, nor tourniquets, nor all the strange appliances of surgery could prevent the patient bleeding to death. We do not yet know for certain what the fibrine is. Chemical analysis shows it to be very like albumen in composition, but then albumen cannot spontaneously convert itself into a substance which has many of the features of a tissue. Some assert that it is a refuse material, resulting from the decay of certain structures. It is much more likely that *it is* albumen on its way to tissue (tendon for example), which, when drawn from the vessels, exhibits to the best of its intrinsic ability a tendency to form some structure, by giving rise to the net-work I have alluded to. How it is kept in solution in the body is also a mystery. Some say it is by the action of ammonia; others that it is due to its motion and contact with living animal membrane; and some consider that as fast as it is formed it is made into tissue, and that what we get in blood is only that which was developed the moment previously.

However, no matter how true or incorrect these views may be, one thing is certain; living blood *does* coagulate. You may say, oh, what matter whether it does or not? No fact in science is unimportant, and this more particularly. It is of the greatest possible value in a medico-legal point of view. For example, a man is found lying on the highway with his throat cut; circumstances indicate that he has arrived at his death by other means, and that the incision in the neck has been made as a ruse to throw the public off their guard. The case then turns on this question, how can you determine whether the wound was inflicted before death or after? The physiologist replies: if before death, the blood will be clotted; but if after, there will be no coagulation. There being no clot detected, the conclusion follows that death was caused in some other manner than that indicated by the injury to the throat.

The blood of a man 10 stone heavy weighs 18lbs., and contains in 1,000 parts
- 795 of water
- 150 ,, corpuscles
- 40 ,, albumen
- 8 ,, salts
- 5 ,, fatty matter
- 2 ,, fibrine

1,000

CAUSE OF THE COLOUR OF BLOOD.

Besides the constituents enumerated in this table, there are, in small quantity, round granular vesicles, larger than the red corpuscles, and which have, from their pale whitish colour, received the distinctive appellation of white blood-globules. These are the parents of the red ones, which, by the way, they set free in dying. It is a curious thing that a fair mother should have a coloured offspring, yet so it is in this case.

The red globules are found by chemists to contain a large quantity of iron; and it is really by this iron that the scarlet tint of the blood is caused. Baron Liebig has given satisfactory proof of this. It is true his opinion has been assailed furiously, but I think that his opponents have had the worst of the battle. I believe his ideas on this question are in the main correct; though I am not prepared to be the champion of their details.

I conceive that the iron is in combination with the white of egg substance, which also enters into the composition of the corpuscles; and that the two together form a new compound consisting of albumen, iron, and soda—not merely mixed, but combined to form a substance differing from any of the three. In the same manner as the letters *o d g*, if put loosely together, as in odg, form no word; but if arranged as *d o g*, or *G o d*, two very distinct sounds, differing from the three letters, and not mere mixtures, result.

Let us see why this is probable. Blood is not always red, it is only so *as a rule* in the arteries; for in the veins it assumes a bluish purple hue.* The origin of this change of colour I shall more fully explain, under the head of respiration; here I can only say that the venous blood carries in it a great quantity of that foul air called carbonic acid. This latter has been produced in the wear and tear of the tissues, and is believed to be the cause of the alteration in the colour of the vital fluid as it flows through the veins.

If we pass through a portion of blood, a current of carbonic acid gas, it is at once deprived of its scarlet hue, and becomes of a bluish purple; if, on the other hand, we force through it a stream of fresh air its original colour is restored. Now, please don't be impatient; we are coming to the point! We have taken albumen, soda, and iron, and made, by artificial means, an albuminate of iron. In the first place we pass through this, carbonic acid, and it assumes a greenish red colour; and in the next we expose it to the action of

---

* A celebrated French physiologist has recently stated that the veins do not always contain dark-coloured blood; that when the vein is conveying the fluid from a muscle which is being exercised, then the blood is purplish, but when the muscle is quiescent then it is scarlet. Also that when the blood is carried away by the veins from a gland (spittle gland, for example) *which is forming its secretion*, it is usually *red*, and *vice versâ*. This is for the following reason · the venous blood from the exerted muscle contains worn-out matter; that from the muscle at rest, none. The venous fluid from the inactive gland contains a great deal of effete matter; that from the secreting gland not so much.

E

the atmosphere, and find it has received a ruddy hue; therefore, we conclude—and I think not unfairly—that the change of colour in the blood is due to the influence of carbonic acid upon the albuminate of iron, which it contains.

The opposite school of physiologists asserts that the cause is a physical one, and not chemical; that in the venous blood the corpuscles become swollen, and, by breaking the light which traverses them, appear dark, whilst in arterial blood they shrivel up, and appear lighter in colour. This is a mere supposition, unsupported by *actual observation*. Both sides agree in this: that their office is to carry the pure air (oxygen) *from* the lungs to the different portions of the body, and bring away *to* the lungs from the distant regions, the effete gas (carbonic acid), resulting from the decay of the system.

It is thought that the blood corpuscles live but for a very short period, and it has been computed that at every beat of the heart *twenty millions* of these little organisms cease to exist. How few think that there are bells within their bosoms tolling the death-knell of so many living beings! Verily "There are more things in heaven and earth than are dreamt of in your philosophy, Horatio." There is no subject in physiology more interesting, or about which there has been more confusion than that of the blood. It has engaged the attention of philosophers from a very early period, and many and curious have been their speculations regarding the nature and circulation of this fluid.

Ere we plunge into the "*medias res*" of the circulation, we must glance at the system of tubes through which the blood is conveyed. These are: Firstly, arteries, or vessels, which carry the blood *from* the heart to the furthest portions of the frame (fig. 27); secondly, the microscopic channels which exist in every part of the body—the capillaries*—ending on one side in the arteries, and on the other in (thirdly) the veins, which bring the blood from the tissues *to* the heart, and so complete the circle. The arteries are quite elastic, being in chief part composed of a material whose properties are similar to those of india-rubber. The capillaries have no distinct coats (?), being only passages tunneled in the soft parts. The veins are not elastic, and are, in addition, provided with valves (like those of a pump), which open *towards* the heart.

The system of tubing just mentioned starts from the great central reservoir—the heart—which is not only the cistern from which the supply to the different parts of the body takes its origin, but also the centre of the propelling power, by which the blood is driven through the arteries. The heart, then, holds the same relation to the arteries that the steam-engine does to the water-piping of our large towns—it pumps the fluid on through the various branches.

---

* From the Latin word *Capilla* (a hair)—a very improper designation, for the vessels thus called are, some of them, so minute, that hundreds of them placed side by side would not be as large as a single hair.

THE BLOOD-VESSELS.

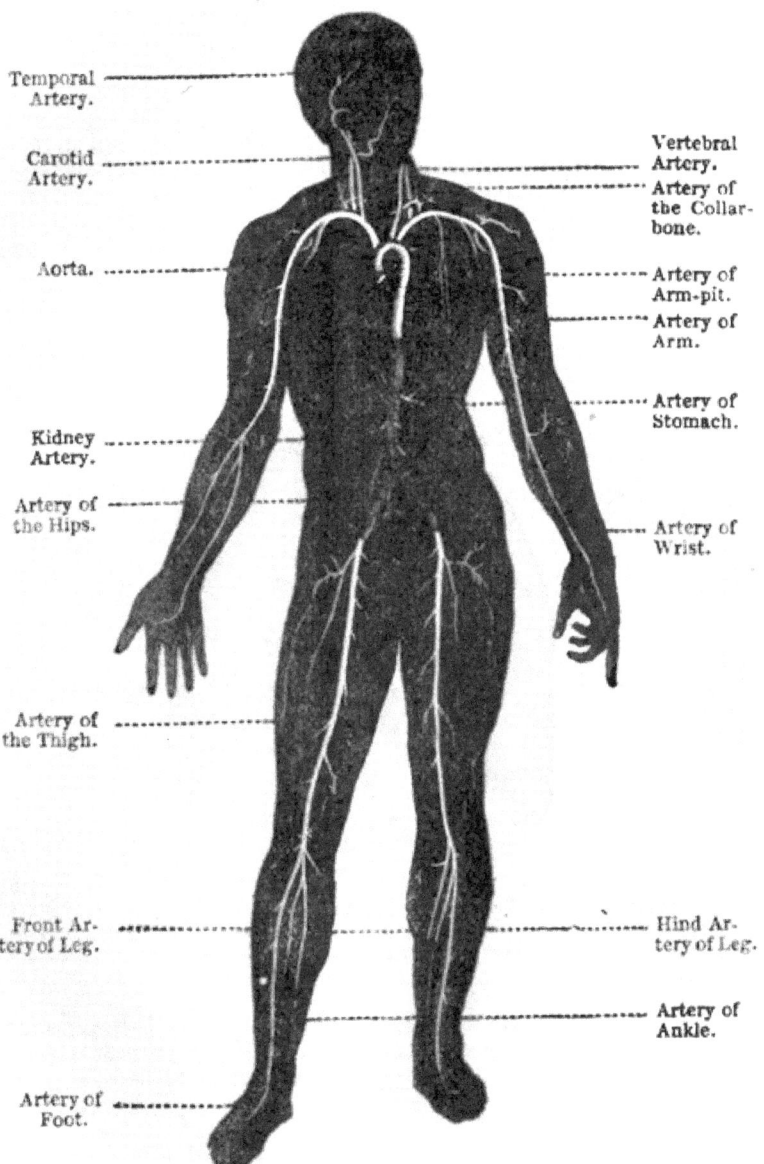

Fig. 27.—Arterial System in Man.

What is meant by the expression "circulation of the blood?" Simply, that the fluid moves in a sort of circle—that is to say, it leaves a certain locality, travels round the body, and returns to the spot from which it set out. Let us say the blood leaves the heart; it is expelled from this organ through the first branch or aorta, and afterwards through its numerous divisions and sub-divisions, till it reaches some distant part—the foot, for example. Here it passes from the arteries into the capillaries (these perforate every portion of the skin, muscles, &c.), and undergoes many changes, giving up the pure materials it contains, to the tissue, in order to repair it, and receiving in return the used up substances resulting from decay. It next enters the veins, and, from the fact of its being impregnated with worn-out tissue (carbonic

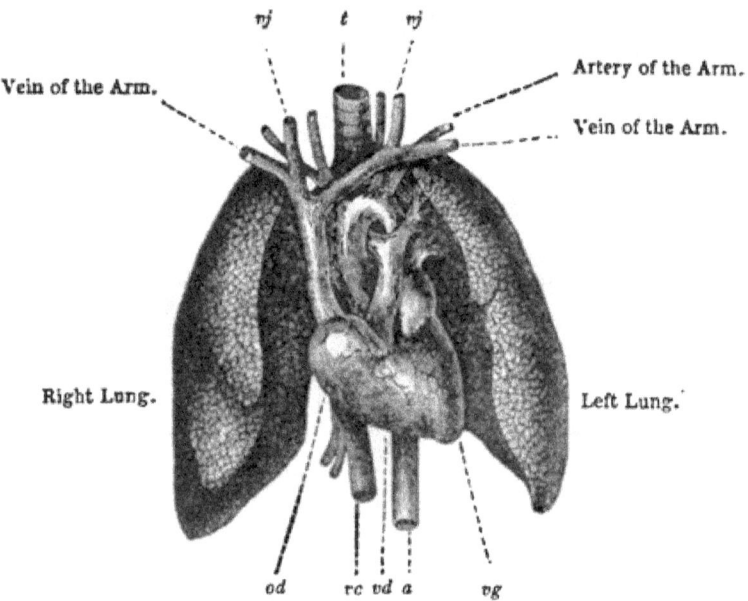

Fig. 28.—Lungs, Heart, and Principal Vessels in Man.—*vj, vj*, the two Jugular Veins; *t*, the Windpipe; *od*, the Right Auricle; *vd*, the Right Ventricle; *vg*, the Left Ventricle; *a*, the Aorta; *vc*, the great Vein which brings impure blood from the Abdomen.

acid, and such like), assumes the purplish tint. Next, it travels along these—flowing from the smaller canals into the larger ones, until at last it reaches the heart, from which it started. This is a rough outline of the circulation; but the complete one is a little more complex.

The heart is a somewhat conical mass of flesh (fig. 28), lying in the chest between the two lungs. From its broad end the great arteries arise, and its point is turned towards the left side, and during pulsation strikes rather forcibly against the flesh between the fifth and sixth

THE CAVITIES OF THE HEART.

rib, or just below the nipple. It is popularly represented as being of the same form as that painted on playing cards; this is a great error, and I should very much like to know who originated it. If, however, you would wish to get an idea of the form of the human heart, examine that of the sheep, which is similar in character. The heart is not a simple bag; it is, on the contrary, a most complicated apparatus, having four distinct cavities, whose openings of communication are provided with valves, admirably fitted to prevent the *return* of blood from one to the other. It is better to say there are two pairs of cavities than four distinct compartments, for in each pair the two cavities are like those of the opposite side. These chambers are termed the auricles and ventricles of each side respectively.

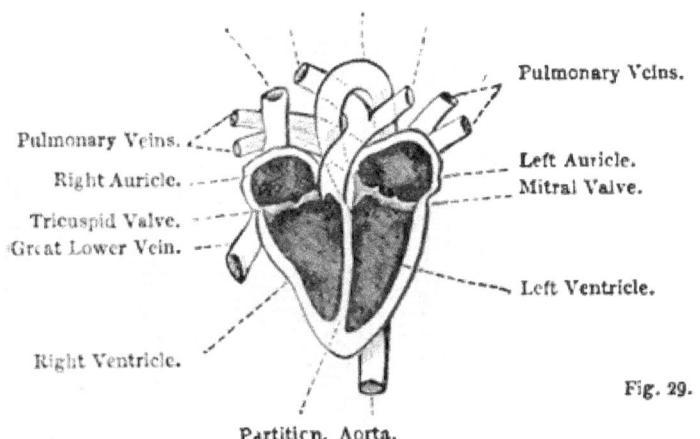

Great Upper Vein. Art. Pulm. Aorta. Art. Pulm.

Pulmonary Veins.

Pulmonary Veins.
Right Auricle.
Tricuspid Valve.
Great Lower Vein.

Left Auricle.
Mitral Valve.

Left Ventricle.

Right Ventricle.

Partition. Aorta.

Fig. 29.

It seems as if the heart was composed of two organs, which are fused together; and when describing the course of the blood it renders the account simpler to speak of two hearts — the right and left. Each heart (or more correctly each side of the heart) is divided into two chambers—an upper and lower—the first receives the blood from the great vein, or veins, and pours it into the second, which then forces it away through an artery (fig. 29). To anyone who has studied hydraulics an objection here presents itself, viz., when the lower chamber is expelling the blood, why does not the liquid flow back into the upper one as well as into the artery? For this reason; the orifice through which the blood flows is guarded by a beautiful membranous valve, which opens into the lower cavity; but when this latter closes on the fluid, in forcing it forwards the valve shuts, and so the blood reaches the artery, having been prevented passing backwards. The upper chamber is called an *auricle*, the lower a *ventricle*; and since there are, as it were, two hearts, it follows that there are two auricles and two ventricles.

Each auricle communicates with a ventricle, but neither the auricles nor ventricles open into each other. Before going into the subject of the circulation I must tell you that the left side of the heart (left auricle and ventricle) always contains pure red blood, and the right side venous or impure. Now, having studied the diagram (fig. 30), you can follow me as I trace the course of the blood from the heart through the various

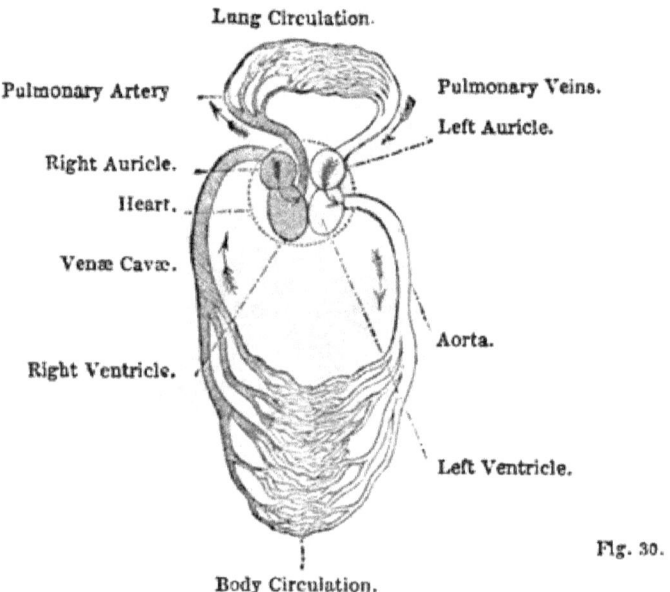

Fig. 30.

portions of the body to the point from which I start. Suppose the blood just poured into the left ventricle; this contracts,* and the contained fluid is now expelled—by a force-pump action—into the greatest of the arteries—the aorta. From this it flows into the various branches, large and small, until at length it reaches the head, brain, arms, belly, legs, skin, and so forth ; thence it is borne into

---

* The whole chamber becomes very much smaller ; this is due to the shortening of the muscular fibres, of which it consists. It is not a mere *india-rubber ball* action, but a special phenomenon, characteristic of flesh. Immediately after the contraction the fibres relax, and the cavity is as large as it was before.

the capillaries, and so comes into contact with the minutest particles of the tissues. From these it is now re-collected into the veins, and eventually is carried by the two venous vessels—the upper and lower venæ cavæ—to the right side of the heart, and emptied into the right auricle. This contracts, and the liquid mass is sent into the right ventricle, which immediately contracts also, and the blood being shut off by the valve (Tricuspid, or three-toothed valve) behind, flows into a large vessel called the pulmonary artery. (*Vide* diagram, fig. 30.) You must not be puzzled to hear this called an artery; for, though it contains impure blood, yet, as it carries blood *from* the heart, it is so named.

This artery is called pulmonary, because it bears the fluid to the lung, the Latin name of which is *pulmo*. Well, then, the blood is driven along this canal to the lungs. Here it is transmitted to the capillaries of these organs, exposed to the air taken in as the breath,

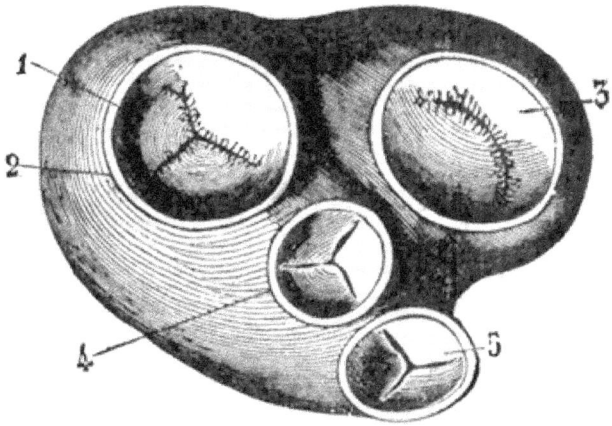

Fig. 31.—Upper surface of the Heart, the Auricles having been dissected away. 1, 2. The Tricuspid Valve; 3. The Mitral Valve; 4 and 5. The Semilunar or Half-moon shaped Valves.

and purified by absorbing fresh oxygen, and giving off the foul air or carbonic acid. But why does it flow upward to the lung and not fall back upon the heart? For this reason: the opening into the heart is protected by three half-moon shaped (semi-lunar) valves (fig. 31), which open in the direction of the lungs, and as soon as the blood tries to re-flow into the ventricles these shut back, and in this way by contraction after contraction of the heart the fluid is driven to the pulmonary organs. Having undergone exposure to the atmosphere and been changed to pure blood in the way I have indicated, it is collected by a number of vessels, which, by uniting with each other, form a single trunk that terminates in the *left* auricle. This trunk is named the pulmonary vein,—*pulmonary* because in connection with the lung, and *vein* because it conveys blood (although *not* of the venous type) *to* the heart.

From this canal it is gradually poured into the left *auricle*, which now contracting, sends it into the left *ventricle* through the opening between the two. The aperture is provided with a pair of valves, that from their resemblance to a bishop's mitre have been called the mitral valves (fig. 32). These open into the ventricle, thus admitting the blood, but when the latter chamber contracts they close, or fall together, and so the blood is driven into the aorta.

Again, the difficulty suggests itself as to its reflux into the cavity, and again, I must say, that here are also three moon-shaped valves, which when the blood is forcibly driven into the aorta lie back against its walls; but when the fluid tries to retrace its steps, shut down—exactly like the trap-door in the pantomime, through which harlequin has been rather forcibly ejected—and thus prevent its readmission into the chamber. The blood has now arrived at the starting point, and so the circulation has been completed. The action of the heart is exactly that of a pair of double force-pumps, as may be seen by looking at the diagram beneath,

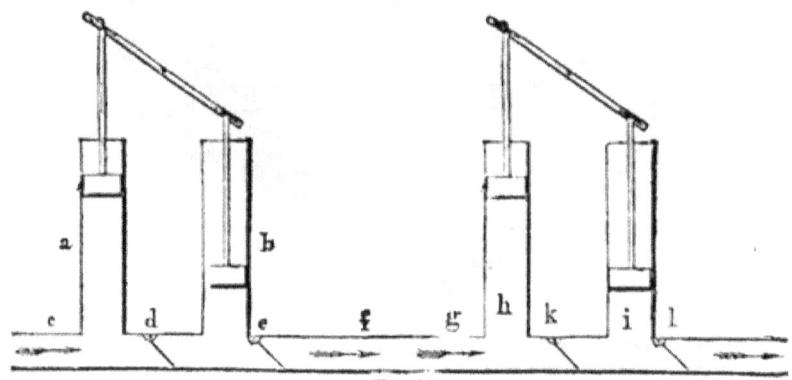

Fig. 32.

where (*a*) represents the right auricle, (*b*) the right ventricle, (*d*) the tricuspid valves, (*e*) the semi-lunar valves of the pulmonary artery, (*f*) the lungs, (*g*) the pulmonary vein, (*h*) the left auricle, (*i*) the left ventricle, (*k*) the mitral valve, and (*l*) the semi-lunar valves of the aorta. The pistons and cylinders stand for the muscular walls of the different cavities, and explain their effect in contracting on the fluid. The chambers of the heart not only contract and so diminish in bulk, but they afterwards expand to their original size. By the first process they expel the blood contained in them; by the second they suck in the blood of the adjoining cavities. That they not only contract but expand also is proved by this fact, that if the heart when contracted be grasped in the hand, the expansion, which afterwards ensues, is very perceptible. The blood flows in a continued rapid stream through the arteries, loses its force in the capillaries, and returns slowly through the veins.

At first one would suppose that as the heart gives a series of beats or contractions one after the other, so should the blood be conveyed through the arterial tubes, in a jet-like manner, by a number of jerks; but this is not so. Why? Why is it not so in the fire-engine, in which, as every one has observed, the force is applied at intervals, by the working up and down of the handles? Because there is a quantity of compressed air, which—being very elastic—when the force ceases, begins to press on the water and so keeps up the flow till the next stroke of the handle. Similarly in the arteries, save that in these the office of the compressed air is performed by their elastic sides.

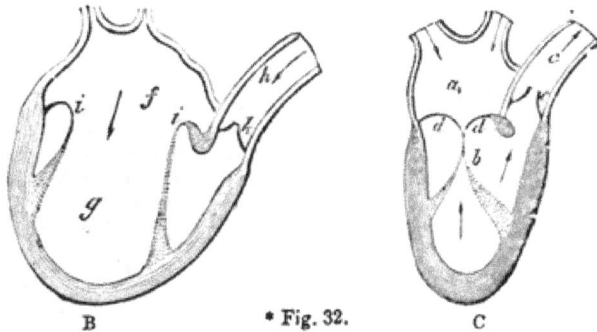

Fig. 32*.—Diagrams illustrating the relative changes in form of Auricle and Ventricle. B. Auricle contracting and pouring blood into Ventricle; f, the Auricle; g, Ventricle; i, i, Mitral Valves open; k, Semilunar Valves closed; h, Aorta. c. Ventricle contracting and forcing Blood into Aorta; a, Auricle; b, Ventricle; d, d, Mitral Valves closed; e, Semilunar Valves open; c, Aorta.—The arrows indicate the direction of the currents.

These, when the blood is first driven into them, are distended, and in recovering their proper position press on the fluid till the next beat of the heart, and so on.

Let us now investigate the forces employed in carrying on this circulation; for that forces there are, is evident. If there was fastened to the open end of the aorta a tube containing 13lb. of mercury, this latter would be supported by the mere pressure of the blood against it. The experiment has not been performed upon man; but from the result of similar experiments upon other animals the deduction has been drawn. The action of the heart, whose equivalent is 13 lb., is certainly the most powerful of all the forces operating on the blood, and is produced by the muscular *contraction* of the left ventricle.

When the column of fluid is thus injected into the arteries these larger branches, by virtue of their elastic coats, equalize the flow; and the smaller ones, owing to their being in part composed of muscular tissue, contract upon the liquid, in this way causing the continuation of the stream. At last, the blood reaches the capillaries, but here, being divided into innumerable minute currents, and actual friction being produced, the accumulated force is expended, and did not some new power exert itself, the circulation would come to a stand still. The new force developed in this locality is that of *affinity*. The tissue toward which the blood is flowing requires certain elements contained in the nutrient liquid. These it robs from the first portions; and those behind having then a greater attraction for the tissue, push the plundered ones before them.

The injured victim is forsaken for one possessing the very attractions that *it* had formerly exhibited. The diagram beneath will make these remarks appreciable (fig. 33).

The globular particles are portions of blood, containing muscle food (*m*) and nerve food (*n*); these are drawn towards the muscle because of an attraction between the two. This next deprives them of *m*; and having lost their attraction, they are driven on by the particles behind, which have not yet lost their virtue. In addition, they are still attracted towards the nerve, on account of their having the substance it demands, and as they are deprived of this also, they are again urged forward by those in the rear. Just as the man subjected to the trying temptations of the world loses first one good trait and then another, till he eventually becomes so vitiated that either death or reform must follow soon.

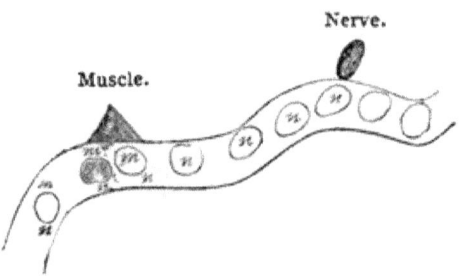

Fig. 33.

By a series of alterations, such as I have pointed out, the blood having lost its nutritious materials, and being highly impregnated with the waste constituents of the tissues which it has taken in exchange, is at length poured into the smallest branches of the veins. These, as I stated, are in general provided with valves opening toward the heart, and it is of the utmost importance that such structures do exist, for the circulation of the fluid through the venous channels is dependent in a great measure upon *them*.

The veins are usually close to the surface, being placed between the muscles and skin; and having exceedingly thin walls, are easily effected by pressure. Now, if we press against a flexible tube containing liquid, this must flow away toward either end; but if it be prevented flowing in one direction, it will *certainly* do so in the other. This is what takes place in the veins—the muscles during the ordinary movements of the body press upon the venous vessels. The fluid contained in these tends to flow in both directions, up and down; but the valves prevent its passage downward (fig. 35), and therefore it ascends, and in doing so opens the valves next

ATTRACTION OF TISSUES FOR BLOOD.

in order, is caught there, pressed on again and ascends in consequence; and this is continued till it has reached the main branches of the system.

The main trunks are two in number; one for the superior, and the other for the inferior division of the body. These have no valves, and hence it is necessary that some compensating power be called into play. This we find in the expansion or dilatation of the heart, which, by a species of suction, draws the venous fluid to the right auricle. You will perceive that we have not yet completed the circle. Having set out from the left side, we have travelled round to the right; but we must go a few steps further. The blood is now sent by the right ventricle to the lungs, passing on its way thither through the pulmonary artery. Here the vital fluid enters another capillary apparatus before it returns to the left auricle and ventricle, and is subject to the same conditions as those we saw when speaking of the capillaries of the body generally. It contains foul gas (carbonic acid), which has an attraction for the air cavities of the pulmonary organs. This foul gas it now loses —letting it escape in the breath—and then it is pushed on by the blood behind, as I explained above, till at length it arrives at the left side of the heart.

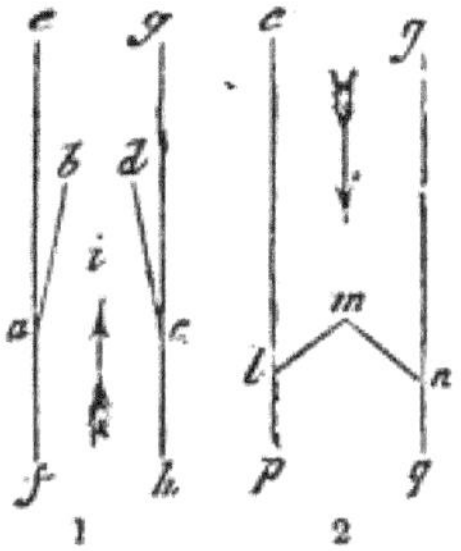

Fig. 34.—Diagram illustrating the action of the Valves of Veins. 1. With the valves open, to allow of the blood passing toward the heart; a, b, and c, d, the valves; e, f, and g, h, the sides of the vessel; i, the central channel. 2. With the valves closed; e, p, and g, g, the sides of the vein, b, m, and n, m, the valves.—The arrows indicate the direction of the currents.

I am afraid you will find the foregone remarks somewhat dry; but I imagine that if they were more extended, it would be only to intensify the puzzle; and unless I made the description longer, it would be difficult to render it clearer.

We are accustomed to feel a man's pulse in order to ascertain the state of his heart; but I am inclined to think we do not all understand the reason, and that many people believe that it is necessary to feel the wrist in particular, and would no doubt ridicule the idea of placing the hand on a patient's neck, or his ankle. The pulse is simply the effect produced on an artery by the force with which the blood is driven through it. The vessel is elongated— stretched out, so to speak—and in consequence lifted a little from its bed, as shown in the diagram (fig. 36). This

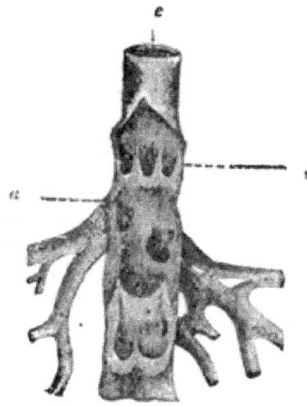

Fig. 35.—Vein laid open; e, end nearest Heart; r, Valve; a, aperture of smaller Vein.

elevation the physician recognises by the pressure against the forefinger, which he has placed upon the skin over the vessel. The pulsation, or pulse, as it is called, may be felt in any artery, provided it be not too deeply situated.

But how does the pulse indicate the state of the heart? It does so, as the water from a pump points to the condition of this latter. If the pump is not being worked, there is no flow from the spout; if the heart be not acting, there is no pulse. If the pump be wrought violently, the rapidity of the current from the spout is sufficiently indicative of this; so, if the heart be contracting rapidly—palpitating—the frequency of the pulsations points *this* out also.

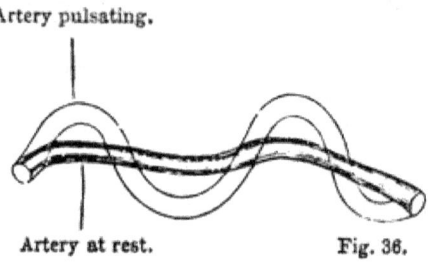

Fig. 36.

Again, to complete our analogy—If the sucker of the pump be in bad repair, the water is heard gurgling backward, and the stream from the spout is of an irregular and imperfect character; similarly in the heart, if the semi-lunar valves are disordered, the blood (by the aid of the stethoscope) is heard trickling back into the ventricle, and the pulse is faint and of a jerking nature.

The beats of the pulse are not so numerous in one individual as in another. Neither are they as frequent in the aged as they are in the adult, nor in the latter as they are in children. Many circumstances affect the pulse, by acting on the heart in the first instance. Of these, muscular exertion and mental emotion are the most important. That the first influences the pulse, any one may prove for himself in the following manner. When lying down, let him examine the rapidity of his pulse; on sitting up he will find it about six beats faster, and on standing erect it will be increased by an additional ten. Or let him contrast the number of pulsations during rest with that during great muscular exercise, and the result will satisfy him. Excitement of the mind accelerates the pulse often to an alarming extent. Which of us has not in some moment of intense anxiety felt the heart thumping against the chest as though it were a coal hammer? Feel the pulse of that man on his trial for murder, who is anxiously scrutinizing the countenances of his jurymen as they return into court with their verdict! Lay your hand upon your sister's wrist as the postman gives his sturdy summons on St. Valentine's morning! Confess, dear reader, that you yourself have at one time had your heart influenced by the "grand passion," and admit, what I want you to admit—the control of the mind over the organs of circulation.

This influence is so familiar to medical men that, knowing the excitement which their presence in the sick-chamber causes, they invariably enter into a quiet conversation with the patient, in order to restore him to his ordinary state, before they examine the pulse. This is not the only way in which the mind operates on the heart; and occasionally, instead of accelerating the pulse, it so affects the system as to paralyze the heart for a time. We all know the result which follows the communication of some dreadful piece of intelligence. The heart ceases to perform its office, and the individual falls as though dead—having fainted. It has been said that the heart is actually ruptured in some cases, and Shakespeare, in conveying an idea of Cæsar's sense of Brutus's ingratitude, says—"Then burst his mighty heart." It is said, also, that the pulse is affected by the digestive organs, being quicker after a meal than before it; and that it is more rapid in the morning than at night; and in the female than the male. The actual number of pulsations per minute is as follows:—

In the infant, 130 to 140 in the minute.
„ adult, 70 to 80 „ „
„ aged, 50 to 65 „ „

The blood travels along the arteries at the rate of twelve inches per second, and in the veins a little slower, moving through only eight inches in the same period. The time required for it to traverse the entire vascular system must not, however, be computed from these data, because a very considerable period elapses between its exit from the arteries and entrance into the veins; in other words, whilst passing through the capillaries. But here chemistry has stepped in to our aid, showing us that when a salt—which is easily recognised by certain tests—is introduced into a vein, it is found to travel the entire circuit in the course of *half a minute*, and therefore that the blood which conveyed it has also "gone its rounds" in thirty seconds.

I have now to draw your attention to what are termed the rhythm and sounds of the heart, for these are matters of the greatest import, not only to the physiologist, but to the physician, whom they enable to judge of the healthy or diseased condition of the propelling organ. By the rhythm we mean the regular or systematic order observed by the heart in contracting and relaxing, or dilating. By the sounds we understand certain noises, accompanying the contractions, and discoverable by the assistance of the stethoscope.

All portions of the heart do not contract at the same time, but as the two sides of the heart work independently of each other, their action is simultaneous to this extent—while the left side is receiving the pure blood and transmitting it to the body, the right one is receiving the impure blood, and pumping it to the lungs. You can at once perceive how impossible it would be for the two chambers of one side to contract together, when aware that the fluid is sent from one to the other. The blood, let us imagine, has been poured into the two auricles, into the right from the veins of the system generally, into the left from the lungs, and then the changes given below take place.

During a *single* pulsation, which I shall divide into four parts as regards time.

In the

- **1st part.**
  - Right auricle contracts, and forces the blood into the right ventricle,
    and, at the same time,
  - Left auricle contracts, and drives the blood into the left ventricle.

- **2nd and 3rd parts.**
  - Right ventricle contracts, and sends the blood to the lungs,
    and, *simultaneously,*
  - Left ventricle contracts, and pours the blood through the arteries of the body.

- **4th part.**
  - A period of apparent rest, during which the auricles are being filled.

All these operations are performed in the interval between two beats of the pulse at *the wrist*.

What can be more extraordinary than the accuracy exhibited by this complex piece of machinery in discharging its duties, as it does incessantly from the cradle to the grave. Well may man exclaim, "Oh, Lord! I am fearfully and wonderfully made." "How manifold are Thy works! In wisdom hast Thou made them all; the earth is full of Thy riches." To those who would find a difficulty in supposing that all the operations above described are gone through in so short a period, I would quote the words of the illustrious Harvey, who, in recounting these, observes that they are not more incredible than those known to be performed "in that mechanical contrivance which is adapted to fire-arms, where, the trigger being touched, down comes the flint, strikes against the steel, elicits a spark, which, falling among the powder, it is ignited, upon which the flame extends, enters the barrel, causes the explosion, propels the ball, and the mark is attained; all of which incidents, by reason of the celerity with which they happen, seem to take place in the twinkling of an eye." *Now* for the sounds of the heart, which, you are aware, the doctor is listening to when he places that queer trumpet-like instrument (stethoscope) against your chest.

> "We told her we were trying,
>   By the gushing of her blood,
> And the time she took in sighing,
>   To see if she were good."

They are, in number, two,—the first and the second. The first is a sort of dull thumping sound; the second more like a short soft click, which follows immediately after the former one. Several words have been devised by different authors to represent the relative length and softness of these sounds,—such as lub dub, lub dullup, &c. I don't think, however, that they express the impression produced on the ear.

Some years since a Frenchman, M. Groux, travelled through this country, exhibiting himself at medical schools, and excited a good deal of interest at the time. His peculiarity was this: the breast bone was actually split into two portions, and, by exercising the muscles of his arms and chest, he could separate these two, so that a triangular space was left, in which the heart beat against the skin alone, in this way allowing the physician to examine the circulating organ as if it was folded in a towel. Some innocent folk fancied that M. Groux could quietly lay open his chest and show the heart itself, but this was not the case. In common with hosts of others, I had an opportunity of examining this individual, and, judging from the sounds heard, it would be difficult to find any two words which, by pronunciation, would exactly resemble them. I *think* that the syllables "loop-up," though not accurately representing the sounds, will give a fair notion of them; *loop* being the first, prolonged and soft sound; and *up* the second, short and sharp one. You can by a little experience acquire a fair conception of these sounds by placing your ear close to the exposed chest of any of your friends, being careful to put it a little below the left nipple. Another excellent method is this: place the fore-finger of your left hand, pointing upwards, alongside your right ear, and then tip it gently at the top with the fore-finger of the other hand, by which a very perfect imitation of the heart's sounds will be produced.

The cause of these sounds, you will say, is obvious enough. If so, you are wrong; at all events, if it is clear to you, it is not so to scientific men, who are disputing year after year as to the mode of production. In order to simplify the matter as much as possible, let us arrange in tabular form the names of the processes performed during each sound.

During, 1st sound.
{ The two ventricles contract.
The point of heart strikes against the chest.
The valves guarding the openings of auricles into ventricles are shut back.
The blood rushes into the aorta and pulmonary artery. }

During, 2nd sound.
{ The two sets of half-moon-shaped valves shut down with a *click*.
The two auricles contract and pour blood into ventricles.
The two ventricles expand. }

There have been so many questions raised as to which and how much of these actions cause the sounds, that a decision in the present state of science would be impossible. But we may safely say that the first results, in a great measure, from the impulse of the heart against the chest, and the shutting back of the valves (tricuspid and mitral) protecting the entrances of the ventricles. Whilst we can *certainly* assert that the second is to some extent, if not entirely, produced by the flapping back of the half-moon-shaped valves of the aorta and pulmonary artery. In proof of this, a most interesting experiment was tried in Dublin in the year 1834, when the British Association met in that city. A needle, barbed at its end, was thrust into the aorta of a donkey, and the valves, being caught by it, and hooked back, it was observed that the second sound had almost completely vanished, not being detected when sought for with the stethoscope. Afterwards, when the needle was removed, and the parts restored to their proper positions, the second sound was detected nearly as usual.

If you have had the patience to wade through the above account, you are prepared to recognize the immense value of a knowledge of physiology to the physician. The slightest alteration of the natural sounds informs him that something is going wrong, and his knowledge of the cause of each enables him to conclude as to what particular locality is deranged, which his acquaintance with practical physic may then give him the power of repairing. In no case could a medical man, ignorant of physiology, hope to produce the cure of a diseased heart, except by the *merest chance*. As well might one entirely unacquainted with the mechanism of a watch expect to repair it when injured, by poking first at this wheel, and then at that; and in both cases the result would be pretty nearly the same,—absolute destruction of parts hitherto in order, and no improvement whatever in those which had been before diseased.

It may be of interest to you to know that some years since a gentleman—Colonel Townsend, lived, who possessed the power of controlling the actions of his heart and lungs. You will tell me this is quite simple—you can do it yourself. Not so fast. It is quite impossible, at all events improbable, that you have any such ability. Draw in your breath, now, and you are, bloated like the frog in the fable, undergoing your torture with heroic fortitude. The gallant gentleman I have mentioned, however, could really influence his heart much in the same way as you can control the action of your fore-finger, which I fancy I see elevated in an attitude of scepticism; as the Ingoldsby legend goes:—

"The sacristan, he says no word, to indicate a doubt,
But he puts his thumb unto his nose, and draws his fingers out."

Colonel Townsend performed the experiment once in the presence of his physician, who cautioned him strongly against its repetition. Nevertheless, he did again exhibit the control he had over his circulatory organs, and this time more triumphantly than ever, for he so completely suppressed the heart's action, that it never throbbed again. The unlucky individual added another name to the list of scientific martyrs which I hinted at in a previous chapter.

It would not be unfair to say, "Prove this circulation you have been talking so much about." I shall imagine that the mandate has gone forth, and having stated my case, that I am bound in support of it to produce my witnesses; or, as the lawyers say, "go into evidence."

EVIDENCE IN SUPPORT OF THE CIRCULATION.

Q. Your name, sir, is Anthropotomist*?—*A.* Yes. *Q.* Having examined the veins of the body, you have found certain structures termed valves?—*A.* I have. *Q.* What do you consider to be the office of these valves? (Court objects to this question, as the subject is not one upon which an opinion is sought.) Are these valves so situated that the blood in the veins can flow freely in both directions?—*A. Most certainly not.* *Q.* In what direction can the fluid travel?—*A.* Only toward the heart. *Q.* Explain this matter to the jury?—*A.* If I took a vein, *that* travelling from the hand to the shoulder, for example, and by means of a syringe tried to force water *from* the shoulder end *to* the hand, I could not do so on account of the valves; but I could force the fluid *from the hand toward* the shoulder without any difficulty. *Q.* If an artery and vein be wounded, is there any difference as regards the end of each from which the blood is poured?—*A.* There is. In the artery it always flows from the cut end *nearest* the heart, in the vein from the open extremity *furthest* from the heart. *Q.* I am, then, to understand that arteries are always carrying blood from the heart and veins to it?—*A.* Yes. *Q.* But if this continues all through life, it is evident that the blood, ever flowing from the hand toward the heart, must be brought there by the arteries?—*A.* Quite so. *Q.* Can you prove that a communication actually exists?—*A.* Yes. If I place the foot of a live frog under the microscope, and observe it, I shall perceive that the arteries terminate in minute channels—the capillaries—which end on the other side in the veins, and I shall observe the blood passing through these channels into the venous system. (Fig. 37). *Q.* The main artery, from which the other arteries are derived, arises from the left side of the heart?—*A.* Yes. *Q.* And the great veins, which the smaller ones terminate in, open into the right side of the heart?—*A.* Exactly so. *Q.* But there is no communication between the two sides?—A. No. *Q.* How, then, can the blood form a circle? —*A.* I shall explain. From the right ventricle there springs a vessel *into* which the venous fluid must be poured by the contractions of that chamber, and *through* which the blood impure in nature is forwarded to the lungs. From these, when cleansed, it is, by a special vessel, returned to the left side of the heart; and this I can show in the lung of the frog, in which the fluid may be seen coursing from and returning

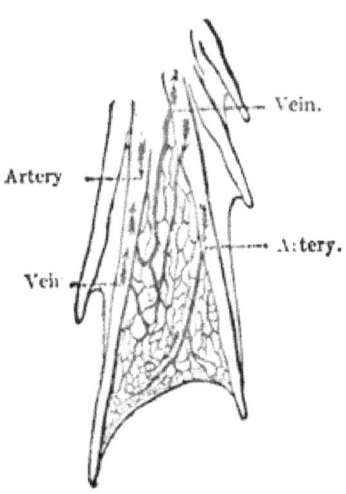

Fig. 37.

---

* From two Greek words signifying "a man" (mortal), and "to cut."

to the circulatory organ. (Cross-examination being declined, the witness has left the box.)

"My lord and gentlemen of the jury:—in this case I am counsel for my client Physiologist, and I shall endeavour, whilst speaking to the evidence which has just been elicited, to deal with the subject as briefly, but withal as clearly, as the circumstances will permit me. It is my intention to show you (as my learned friend observed in opening the case), that a certain individual, one Sanguis by surname, has been during a considerable period of time—viz., the life-time of his master, Homo, in the habit of travelling in a peculiarly circuitous manner along sundry canals (blood-vessels), from and to a certain metropolis—the heart. That the routes he has pursued in going and returning have not been identical; but that, on the contrary, he has invariably set out by one series of canals—the arteries—and come home by a second and distinct water line—the veins. That *you*, gentlemen, may now accurately conceive the course followed by this sanguinary character, I beg to refer you to the diagrams and schematic representations (Figs. 29, 30, 31) already laid before you by my client.

"You are now prepared to follow me as I move forward step by step through the complex chain of circumstantial evidence given by the distinguished man of science you have had before you. You observe in the delineations of the metropolis I have alluded to, it is distinctly shown that the two portions have no direct communication with each other; that certain canals lead from the left side to the various regions around, and that these canals *then* are continuous with others which return to the right side—the first being called arteries, the second veins; and, finally, that adjacent to this metropolis is a great tract, known by the name of Lung, to which a canal passes from the right wing of the metropolis, and *from* which a similar canal returns to the left wing.

"Now, it is admitted that the prisoner was seen in the lung, in both wings of the metropolis, in the various regions lying round, and in both the arterial and venous canals; from which we conclude that he has travelled in a circle. He alleges, in defence, that he has never adopted a circuitous route, but has journeyed from the heart to the surrounding country, and also from the right side of the heart through the veins to the same localities, constantly going and returning by the same canal. To this, gentlemen, I can only say, recall the statements, the *reliable statements* of Anthropotomist. That gentleman informed you 'that a system of turnpikes existed along the venous canals at which passengers were refused passage in any *but a homeward direction;* also, that owing to the existence in the arteries of a very powerful current in the *outward* line, it would be impossible for any individual to return by those canals; and finally, that the only communication between the right and left districts of the metropolis was *viâ* the lungs.' Recall, I repeat, those statements, and I believe that as conscientious and sensible men, viewing the matter in the most impartial and unbiassed manner, you will feel bound to bring in a verdict for my client.

For Sanguis has been seen in both districts of the heart, in the lungs, arteries, and veins; and the only mode by which he could *consecutively* have presented himself in these different localities is by performing the circuitous journey I have contended for."

A wonder it seems, that the old barber-surgeons of England, who from their practice of phlebotomy must have been well acquainted with the difference of direction observable in the flow of blood through arteries and veins, never hit upon the cause, never even *for a moment* lighted upon the real object of this distinction. For be it known to you, dear reader, that the barber-surgeons of this country were once a much more respectable body of men than they are in our days. And when I say in our days, I say it advisedly, because the barber-surgeons, though nearly extinct in Great Britain, are still represented by a few antique specimens, who practise the arts of blood-letting and shaving in certain remote localities.

Why, you will ask, should these professors of the *ars medendi* have struck at the root of this obscurity? Because, from their practice of bleeding, they were familiar with the necessity of putting a bandage on the arm before opening the vein, in order that no air should be sucked or drawn into the end nearest the heart, as soon as the vessel had been divided, and because they knew very well that prior to the application of the lancet, when the bandage had been on for a short time, the artery was swollen *above* it, and the vein *below*. The fellows, it is thought, though *shavers*, were not otherwise observant; and from having been brought up in a state of barberism, much allowance must be made for their want of progress.

Are there many who know the meaning of the barber's sign which projects boom-fashion from the windows of some of the perfumers of "the great unwashed?"—a long pole some nine feet in length, of a red colour, with a white band twining around it from end to end. This is its signification, and one that I fear many of the proprietors are unaware of: the scarlet rod is intended to convey a notion of blood, and the white stripe which winds around it of the bandage employed in the operation of blood-letting.

That the surgical practice of these reputable members of our profession was not confined to the mere abstraction of the vital fluid is shown by the "Diary of Samuel Pepys," where the eccentric author mentions having broken his head whilst returning from some of his not unfrequent symposia, and having had it repaired by a barber-surgeon, for which he "did pay" five shillings. Here I am reminded of the fact that, in the present day, our "general practitioners" have only to undertake the charge of the public hair and whiskers, and they will then pursue an avocation whose duties are vastly more multifarious than those of *their* respected prototypes; inasmuch as they will engage to shave you, bleed you, mix your physic, and cut your hair, "all for the small charge of *two and sixpence*."

In Germany the barber still fills the office of phlebotomist—indeed, this is as much a portion of his trade as the employment of the shears;

and the sign above his establishment is a series of brightly polished brass plates suspended from a horizontal bar, and swinging to and fro by the action of the wind. The plates are significant of the circumstance, that the blood, on flowing from the vein, is usually received in a dish or some such utensil.

I recollect not long ago, being in a German theatre, witnessing the performance of one of those burlesques, of which the Germans are so fond, when suddenly my ears were startled by a shrill scream, arising from the pit (*sitz parterre*). On looking down, immense confusion appeared to have seized upon the people below, which seemed to have been caught up by the *dramatis personæ*; for in less time than it has taken you to read this description, the curtain dropped. Anxious to discover the origin of all this disturbance I descended to the pit, and found that during the performance a man had been attacked by apoplexy; his wife, seeing him fall in a state of insensibility, had shrieked out, and ere the curtain had fallen a prompt and energetic barber, who had been sitting a few seats in the rear, had opened a vein upon the spot, and was now having the patient conveyed to another locality. *Obstupui. Comas* did not *steterunt*, but *vox faucibus hæsit*—which freely translated means, I was dumb-foundered, to think that almost in the twinkling of an eye so many things had happened. I returned to my box seriously impressed with the belief that a corps of German barber-surgeons might be imported into *the country districts of England* with advantage.

Who discovered the circulation of the blood? A very nasty ticklish question, and one not easily replied to. In this age of perverseness and aping after originality, it is often endeavoured to earn a name by writing a work to blacken a character heretofore without blemish, or canonize some wretch whom two centuries of society have condemned. This habit has not been foreign to physiological writers, and, therefore, the question as to priority of discovery, not only in the case of the circulation, but in many other instances also, is a vexed one. If I were sworn to state my opinion regarding the discovery of the circulation, I should say: on the one hand, Harvey does not merit all the credit he has received; on the other, the circulation was not *proved* before his investigations, whilst, in any case, he did not give all the details.

Galen, who lived A.D. 150, was the first to form any true idea concerning the process, for *he* asserted that the arteries carried blood, and not air. Vesalius was the next in the field, and showed (about the middle of the sixteenth century) that no direct communication existed between the right and left sides of the heart. He was a most enthusiastic and philosophic investigator, and suffered under the religious prejudices of the period; for, not contenting himself with the examination of the bodies of the lower animals, he, on one occasion, made, *as he thought*, a "post-mortem" examination of the body of a young nobleman. Judge of his horror and astonishment on finding, when the chest was opened, that the heart was still beating. This became noised abroad, and he, poor man,

HISTORY OF DISCOVERY OF CIRCULATION.

was obliged in penance to undertake a mission to the Holy Land, and afterwards fell into disgrace and temporary oblivion.

The next labourer was Servetus, who, with tolerable distinctness, pointed out that the blood was carried *from* the heart *to* the lungs, and then returned; but *he* also came to an untimely end. He was a divine; and, indeed, published this physiological discovery in a theological work; and, to quote the words of Mr. Lewes, "both he and his treatise were roasted by Calvin." A few years subsequently, Cæsalpinus, a distinguished botanist, first coined the expression, "circulation of the blood," and described the circulation thus: "In animals we perceive the food brought by the veins to the heart, ... and it is distributed over the entire body by the *arteries*." Next, we observe that, in 1574, Harvey's own master, Fabricius, discovered the valves in the veins. Finally, in 1619, we find Harvey himself appearing in the arena. It is quite certain that much had been done for him. Thus:—

1st. It had been proved (?) that the arteries carried the blood *to* the body.
2nd. That there was no communication between the two sides of the heart.
3rd. That the blood was sent from the heart to the lungs, and returned to the former.
4th. That the food was brought by veins to the heart; and—
5th. The term "*circulation* of the blood" had been employed.

But still, all the information was in crude form, and no decided conception of the *entire* course of the blood existed. All these deficiences he supplied, and to him *all* the merit of *establishing* the "circulation" is due.

Harvey, however, was unaware of the manner in which the fluid passed from arteries into veins, and this hiatus remained unfilled till the discovery of the capillaries by Malpighi, in 1661—a discovery which completed the matter, and which is a proof of the immense advantages accruing to physiology from the use of the microscope.

CHAPTER VIII.

Respiration—What we mean by Drowning—The Chest—Lungs—Air Vesicles—Suction—How the Air is introduced into the Air Cells—Enlargement of the Cavity of the Chest—Ascent of Ribs, Descent of Diaphragm—Ordinary Respiration—Elasticity of the Lungs—Resistance to be overcome in Inspiration—Effect of a Gun-shot Wound in the Chest—Law of Mutual Diffusion of Gases—Intermixture of Hydrogen Gas and Carbonic Acid—Capacity of the Lungs—Spirometers—What we inhale and exhale—Experiment—Is Carbonic Acid *generated* in the Lungs?—Experiments on Birds—Composition of the Air—Poisonous Action of Carbonic Acid—Importance of an attention to the Principles of Hygiene—Why we breathe—Organ of Voice—The Larynx—The Vocal Chords—Action of these latter—Use of Epiglottis—Voice of celebrated Singers—Speech—Stammering—Ventriloquism—Whistling—Sighing—Yawning—Sneezing—Laughing—Consumption and Bronchitis.

A MAN tumbles into the river, and you say he's drowned. What do you mean? Why, that he is suffocated—asphyxiated—you know. Ah, dear reader, you cannot thrust those long words down my throat. Confess your ignorance, and say that all you know *is* that he is drowned. His breathing ceased because he had not air; his blood, not being purified, would not pass to the left side of the heart; and so the circulation was arrested, and, in consequence, life itself. Now the solution of all the questions involved in these processes is only to be arrived at by the study of the function of respiration. This study we shall pursue in the same way as we followed that of digestion and circulation; viz., we shall first examine the machine; next, show how it works; and lastly, treat of the advantage of this labour.

The organs engaged in respiration are, the lungs, windpipe, and chest. The lungs, or lights as they are called by butchers, are two large, spongy masses, which, during life, fill the whole cavity of the chest, just allowing the heart and great blood vessels to rest between them, and the gullet to pass down at the rear (*vide* fig. 38). They are enclosed in two membranous sacs, and though enclosed are yet outside them. You'll say this is a paradox; how can they be outside and in? Have you ever seen one of those glasses with hollow sides, partly filled with some coloured liquid resembling porter, and which amuse children exceedingly, from its being impossible to spill the supposed contents? If you have, you can understand how, if you pour water into a glass of that kind, it will virtually be *in it* and yet *not* in it. Or, suppose your own head buried in a double nightcap; here, as you know, your head is in the cap and yet *not* within it. Well then, in the same way do the lungs lie in their membranous bags. The real interior of these is unceasingly being oiled by a peculiar secretion, so that in this manner the two surfaces, though constantly rubbing against each other, sustain no injury, owing to the friction being provided against by the oiled surfaces. These sacs are termed *pleuræ*,* and when by any cause their lubricating power ceases, the rubbing of the rough sides causes intense agony in drawing the breath, in fact, the bags are diseased, and the malady is named pleuritis, or in common parlance, pleurisy.

The lungs consist, like other glands, of a canal system (fig. 39, a, b, c, e), by which their secretion (carbonic acid) is carried out, and a quantity of vesicles, by which it is separated from the blood. The tubes which convey from the lungs the foul air and vapour (which are *their* secretions) commence at the windpipe. This, in its turn, begins at the larynx (that group of structures of which the Adam's apple is one), and, after descending into the chest, divides into two smaller conduits—the bronchial tubes — which again divide into two lesser ones, and so forth; this process of division and subdivision being carried on till eventually an arborescent collection results, whose final branches are microscopically minute. If we examine the furthest end of one of these lungs we shall perceive that it is connected with an irregular labyrinth of cavities, which is named a lobule. Now this lobule is a portion of the vesicular part of the lung, and is composed of an

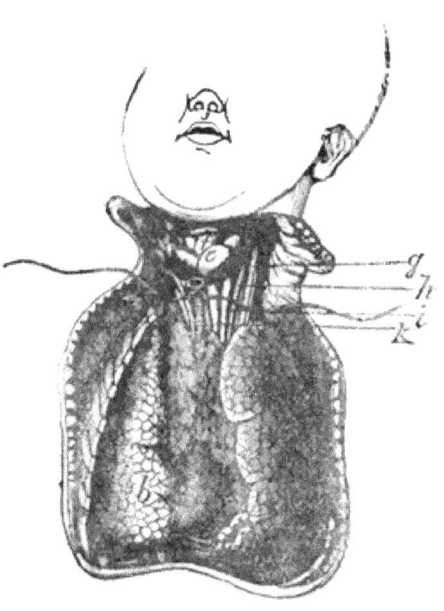

Fig. 38.—Chest of a Child, the heart, bone, and front of the ribs having been removed. The lungs have been inflated, and the windpipe has been tied; a, the heart; b and c, the lungs; d, a large gland. The other letters refer to parts seated in the neck.

immense number (18,000) of cells, about the 150th of an inch wide. These communicate very freely with each other, are bounded by a delicate transparent web, and have, lying between them, *thousands* of exquisitely slender capillary blood-vessels. It has been computed that of these air cells there are in the human lungs not less than *six hundred millions*—a fact well calculated to impress upon us the vast importance of the pulmonary organs.

You have already appreciated the relation of these different parts. The air cells are in clusters, those of each group or lobule being con-

* From the Greek for lung.

tinuous with each other. To each cluster is attached the ultimate branch of a bronchial tube, which, as it travels upwards, unites with others, until at last, through the windpipe, a communication is established between the atmospheric air, externally, and the minute congeries of air cells within. The fine blood-vessels are situate between the cells, being, as it were, packed along with them, so that each capillary tube has an air cell on both its sides.

The larger bronchial tubes are lined on the inside by a beautiful delicate down, soft as velvet, of which I have just placed a portion under the microscope; and what a pretty sight is presented—a field of corn in miniature! This down is formed by an almost infinite number of extremely minute, hair-like filaments, resting on club-shaped projections, and perpetually moving in one direction, giving exactly that appearance to the eye which is produced by a meadow swayed in gentle undulating curves by the action of the wind (*vide* fig. 40). I now drop a small quantity of a solution of potash upon the specimen, and I have a "dissolving view" produced, for the elegant little filaments (cilia) have vanished.

Surely, these exquisite organisms are not without a purpose? There must be some office which they fulfil.

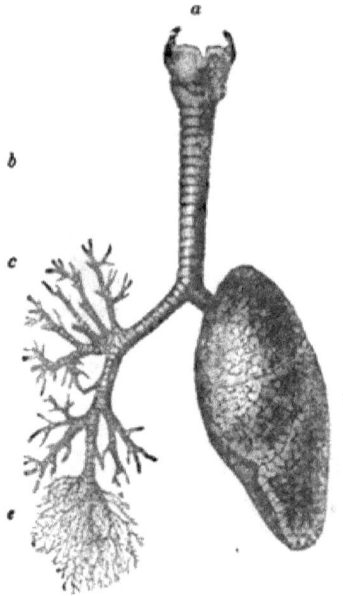

Fig. 39.—Lungs and Windpipe of Man. On the right side the soft tissue has been dissected away, and the tubes alone are seen. *a*, the larynx; *b*, the windpipe; *c* and *e*, large and small bronchial tubes; *d*, the left lung.

"Oh, happy living things! No tongue
   Their beauty might declare:
A spring of love gushed from my heart,
   And I blessed them unaware."

The cilia always move in *one* direction, and in the bronchial tubes this is toward the windpipe—upwards. Hence all particles of dust, all sorts of materials in a finely powdered state, which may be accidentally sucked in during respiration, are prevented descending into and accumulating in the air cells by the influence of these cilia. A small, almost atomic, portion of road dust we often draw into the lungs on a blustry summer's day, but it effects no injury, for it hardly has got in before the cilia "take it in hand," and it is sent back again from one to the other till it has reached the mouth.

Were it not for this grand provision, all the millers and stone cutters would be exterminated in a very short period. Even as it is, they do meet their death sooner than other folk, because of the inability of the cilia to prevent all the particles entering. A more energetic atom than usual will elude their vigilance and slip down occasionally, and this being oft repeated, the collected matter sets inflammation and other morbid processes agoing, which end in the extinction of life. The bronchial tubes and windpipe are composed of a kind of gristle or cartilage, mixed with tissue of a sinewy description; and in addition to these there are a few fibres of muscular tissue (flesh). These muscular filaments can hardly be seen, but a very ingenious experiment has shown us their existence. Muscle always contracts when galvanized, and therefore if a galvanic shock causes the lung tissue to contract, it probably contains muscle. An English physiologist having dissected out the lung and bronchial tubes of an animal, placed the entire organ so that the opening of the windpipe was opposite the flame of a candle; next he applied the wires of the galvanic battery to the lung, and he heard the air rush out, and saw the candle extinguished.

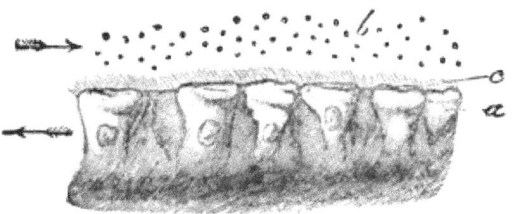

Fig. 40.—A portion of the lining membrane of the Windpipe, showing the Cilia. *a*, the club-shaped cells; *b*, particles of matter; *c*, the Cilia.—The arrows indicate the direction of the currents.

This information is of great value, for it shows us the cause of what many people have experienced—spasmodic asthma. In this disease the chief symptom is a feeling of sudden suffocation, and this is produced by the contraction of some muscular bands surrounding the smaller bronchial tubes, thus closing the channel and preventing the passage of air to the cells. Now, it has been found experimentally that such substances as *thorn-apple* and *deadly nightshade* prevent, to a great extent, these contractions, and therefore we employ both these remedies to alleviate the unpleasant sensations. I merely mention these facts to show you how we obtain a knowledge of the cause of a disease through the assistance of physiology, and that when we wish to discover a remedy, we must also appeal to the same branch of learning.

The lungs then, as you now understand, are a pair of spongy sacs, divided internally into millions of little compartments, which the air reaches after its journey through all the bronchial tubes.

The next points are, how we bring the fresh air in, and force the foul air out. These we can only solve by a reference to the mechanism of the chest. The latter is, as I before said, a boney cage, but though a cage, the air cannot enter it though the framework. (See fig. 41.) The component wires having their intervening spaces filled up by flesh, and the floor being formed by a large dome-shaped muscle, which only allows the gullet and great vessels to pass through it. At the top also it is shut in by flesh and membrane, which allow the gullet and windpipe to enter and the large arteries and veins to make their escape. Therefore, since the windpipe is the only channel that communicates on the one hand with the lungs, and on the other with the atmosphere, it follows that *through it alone* air can enter the chest.

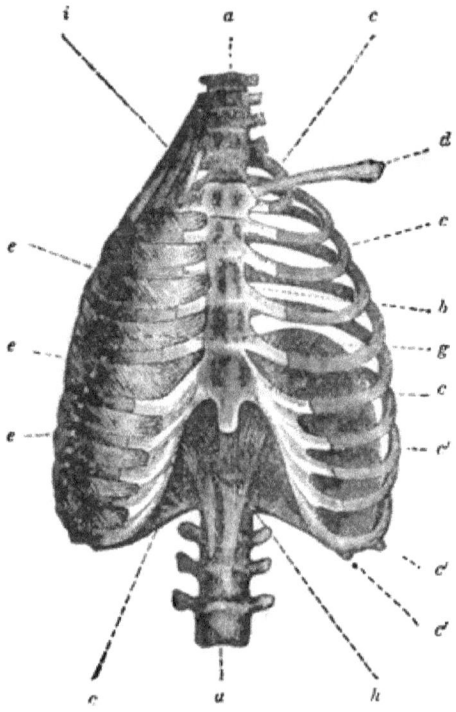

Fig. 41.—Chest of Man.—*a, a*, Backbone; *b*, Breastbone; *c, c*, Ribs; *d*, Collarbone; *e, e*, Muscles connecting Ribs; *i*, Muscles which raise the first and second Ribs; *c′, c′, c′*, the Diaphragm or fleshy partition; *h*, Sinews of the same.—The dotted line represents the position of the Diaphragm on the right side.

But, since the lungs are themselves air-tight, it also follows, that nothing can pass from them into the chest (*i.e., between them and the ribs*), and so we arrive at the conclusion that the chest is a perfectly closed cavity.

Don't, I beg of you, imagine that this description is unnecessary. It would be quite impossible without it to make you realize the processes of inspiration and expiration. Even still, ere you can thoroughly bring the thing home to your minds, it is requisite that I carry you two steps further. First, I must tell you that, if there be placed in an air-tight chamber a bladder which opens outwards, and

WHAT IS MEANT BY VACUUM. 75

that you try to enlarge this cavity, the bladder will be inflated. This is proved experimentally in the following manner:—

Take a hollow cylinder of glass (*a*), fitted above with a cap which is pierced by a tube, to which a bladder (*b*) is firmly tied; let there be at the bottom a piston (*c*) with a short handle, and have the whole apparatus air-tight; then, the bladder will remain as represented in fig. 42. If, however, you now draw down the piston in this way, *apparently* increasing the capacity of the cavity, the bladder will be at once inflated. To one who is entirely unacquainted with pneumatics, this seems very extraordinary. It is not so. The filling of the bladder depends upon the fact, that the atmosphere presses equally in all directions. Thus, when the apparatus was first put up, the air *in* the cylinder pressed against the outside, and that without against the inside of the membrane, with equal power; but, when the piston was depressed, the air in the cylinder became thinner; its power being diminished, the resistance of the outer air overcame it, and so the bladder was blown out till the two pressures *again* balanced. This is exactly what takes place in the chest. The cavity, during inspiration or drawing in the breath, is enlarged, and the pressure of the atmosphere dilates the lungs, filling them with air. I dare fancy you'll tell me, you knew all this before. You were quite aware that the air was carried into the lungs by "suction." But you might as well have said, simply, that *it was* carried in, for the word suction affords no explanation whatever. Torricelli, a great natural philosopher, gave no better idea of the *real* cause, when he said that "Nature abhors a vacuum."

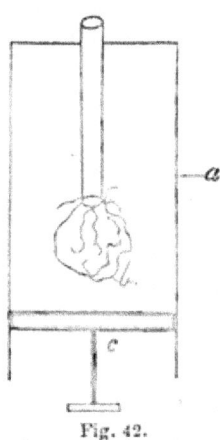

Fig. 42.

We have got the right leg forward now that we know so much; the left will follow whilst we are examining the contrivances by which the enlargement of the chest is effected. Keep one eye on the diagram (fig. 41) while you are following me with the other. Observe the way in which the ribs (*b*) slope downwards from back to breast-bone, and mark also the convex dome formed by the great fleshy floor or diaphragm (*c' c' c'*), and you will agree with me in thinking that when the latter is drawn downwards and the former are elevated the cavity of the chest will be much increased in dimensions. The depression of the diaphragm at once accounts for an enlargement of the chest, for the process is fairly comparable with the descent of the piston in the cylinder; but the expansion arising from the elevation of the ribs is not so evident.

First let me show you how the ribs are elevated. Each rib has a sort of moveable joint situate at the junction with the backbone, and this permits it to work up and down when acted on by the muscles. Every rib is connected with that above and below it by layers of muscular tissue, and the upper one of all, or first rib, as it is called, has attached to it certain band-like muscles which arise from the vertebræ of the neck. (*i*) When the band-like muscles contract, they draw up the first ribs; and the muscular tissue, passing from these to the second, the second to the next, and so on, being contracted at the same time, the whole of the ribs are brought more into the horizontal position, each one describing a part of a circle, of which the centre is the joint at the back. The leather portion of a large smith's bellows when seen at work gives a very good idea of the rising of the ribs; the great board being first elevated by the action of a lever, and then the various folds of leather ascending slowly one after the other. By this elevation of the ribs, the chest increases more in width than in length, because, as the ribs pass from the inclined to the horizontal plane, their front ends become more distant from the backbone. This is seen in the accompanying diagram, where the oblique lines represent the ribs when depressed, and the horizontal ones when elevated (fig. 43). Should you require a more tangible proof of this, I would say, take a pair of parallel rulers, place them in the erect position, and then the oblique connecting rods will stand for the ribs where depressed. Now separate the two bars, let one be back and the other breast-bone, and as soon as the rods are placed horizontally, you will find a considerable space of quadilateral outline, lying between them.

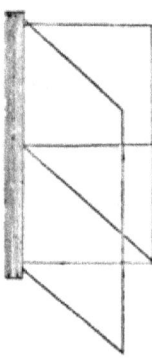

Fig. 43.

The diaphragm is more constantly employed in the labour of respiration than the ribs, and during its descent, presses upon the stomach and guts, driving them forward, as everybody knows by experience. The ribs are elevated only once in every five inspirations, and then "the long breath" is taken. Hitherto I have been speaking of inspiration, but I have not yet touched on expiration, because it is more a *passive action*, if I may use so contradictory an expression. The lungs are exceedingly elastic, and this elasticity must be overcome in their inflation; from which it results, that so soon as the vanquishing power ceases to exert itself, the tendency to return to the natural condition will expel all air which has been introduced by forcible means.

In inspiration the diaphragm contracts and descends, and the rib-muscles also contract and elevate the framework; then, the pressure of the atmosphere exceeds that of the lungs' elasticity, and the air rushes in, but after this the diaphragm relaxes and *ascends*, and the ribs *descend*, and now the lungs of their own intrinsic strength eject the air which had taken "forcible possession"—an act of expiration has been performed.

A man unfamiliar with the object of the balance-wheel in a watch, is given a "movement" in which this mechanism is injured; the injury being evident, he winds up the main-spring, and ere he has removed the key is aware of a buzzing noise, and beholds the hands travelling with fearful rapidity round and round the dial. Instantly the idea strikes him that the balance-wheel is adapted to "regulate the time." So it is in our complex machinery; we ourselves are ignorant of the exact purpose of many beautiful appliances till they are disordered, and then we perceive by the effect produced on the general system, the advantage of what might have previously appeared as a means without an end.

To illustrate the use of the peculiar pneumatic contrivances seen in the chest, we need only look at a man whose lung has been wounded (let us say by a musket-ball). The cavity is no longer air-tight, and respiration is performed with great difficulty, the air being admitted and expelled through the wounded orifice as freely as through the windpipe, or more easily, and suffocation not unfrequently occurring from the inability to form a vacuum, and by that means convey the gases of the atmosphere to the ultimate cells.

Some conception may be formed of the immense elastic power of the lungs *themselves*, when it is known that in the greatest inspiration a resistance equivalent to 150 lbs., or nearly eleven stone, has to be overcome in the male, and about 120 lbs. in the female. This is not all, there is also a great amount of elastic power in the walls of the chest,—so great, that to overcome it a force of 104 lbs. is required for every deep inspiration. Add this to the inherent resisting power of the lungs, and you will have a sum of 254 lbs., or more than 18 stone, a force which the muscular apparatus of the chest exerts whenever you draw in a very long breath.

It seems astounding that a man can, with his ribs and diaphragm, exercise an amount of power equal to that of a coalheaver when he lifts a sack weighing 2¼ cwt. upon his shoulders; yet, that this statement is not fallacious has been demonstrated by experiment. The lungs never expel all the air which they contain after any inspiration, for it is found that the lungs of man can hold 240 cubic inches, and those of woman about one half this, whilst there is little more than 20 cubic inches drawn in and sent out, ordinarily. How then, you will naturally inquire, is the impure air which remains got rid of? To explain this I must digress.

Chemists have discovered that gases invariably mingle with each other in obedience to a certain law termed "*mutual diffusion*." For instance, carbonic acid, or foul air, is much heavier than the atmosphere, so much so that you can decant it from one vessel to another, as you would water, although apparently the vessels are empty. Take a bottle of carbonic acid gas, and invert it over a lighted candle, and the flame will be instantly extinguished. And since this noxious gas is perpetually being formed, in large towns more especially, we should expect that eventually it would, as it were, inundate the low countries, and put an end to life. Indeed it would do so, had not Providence ordained otherwise by this statute of "mutual diffusion," "by which it is enacted" that all gaseous fluids of different weights—bulk for bulk—shall so mix and associate with each other, that destruction of life may be avoided.

The following very interesting experiment shows the mode of execution of this law: Let a glass jar, with a stop-cock, be filled

78   POPULAR PHYSIOLOGY.

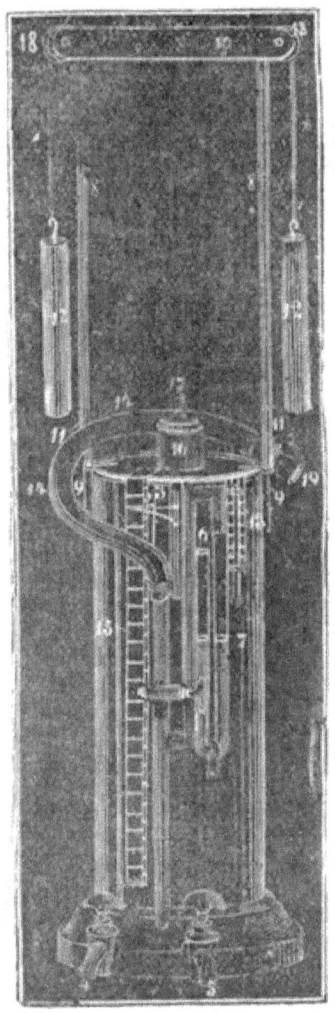 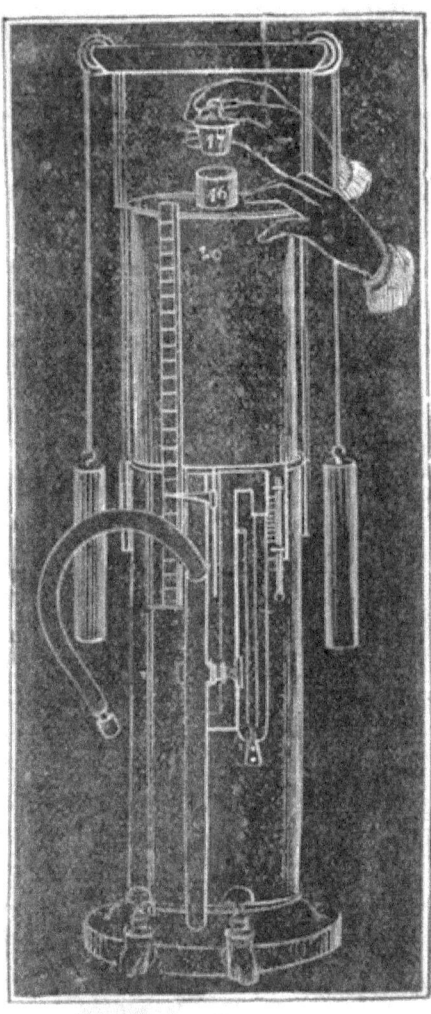

A   B

Fig. 44.—Spirometer, or instrument for measuring the capacity of the Lungs for air. It is a kind of gasometer. When air is blown into the body of the vessel, through the tube, 11, 17, 14, 19, the moveable cylinder is elevated, as shown at B: and as it rises, it marks off on the graduated scale the number of cubic inches of air it has received. 8, 9, 10, 11, framework supporting the weights, 12, 12, which balance the moveable cylinder; 18, 18, pulleys for the cords; 16, aperture through which air is expelled after the instrument has been used; 17, the stopper; 15, the scale which exhibits the amount of air breathed into the spirometer; 13, thermometer to show external temperature; 20, the moveable cylinder elevated in fig. B; 13, tube containing coloured alcohol, to indicate the relative pressures of the air within, and external; 4 and 5, cocks for the removal of water.

PULMONARY INHALATIONS AND EXHALATIONS.

with carbonic acid, and placed in the usual position; then let a second vessel of the same sort be filled with the lightest of all gases —hydrogen, and let this be inverted, the two stop-cocks being united. Next, let the cocks be opened, and the apparatus left to stand for some hours; after which time it will be found by chemical tests that the heavy gas has, to a certain extent, ascended, and is present in the upper jar, and the lighter one has come down, and may be detected in the lower one.

Revenons à nos moutons—or to apply the information we have just arrived at to the function of respiration, I may observe that the same "law of diffusion" is applicable to the gases in the lungs, and that, although but twenty-two cubic inches of air enter these organs at each inspiration, yet there is not left noxious gas in one part whilst the pure fluid circulates through another, but, owing to this principle of mingling, a semi-pure atmosphere is supplied to *all* the minute compartments.

For what purpose is all this complex mechanism? To bring in fresh air, and send out the impure. What is the use of the lung, and of the pure air? And where does the foul air come from? Answering these queries consecutively: the use of the lung is to expose the blood to the oxygen of the air; the object of the air is to purify the blood, and give it oxygen for the tissues; and the foul air arises from the uncleansed blood. You are already conversant with the course of the venous blood; you know that it is pumped by the right ventricle to the lungs, and that here it enters millions of microscopic vessels, which lie between the cavities or cells. In these localities the corpuscles, highly charged with carbonic acid and watery vapour, which they have derived from the tissue-refuse, meet the atmospheric air—an almost inconceivably thin membrane intervening—and a species of *osmose* takes place, the carbonic acid and water deserting the blood disks, and the oxygen being simultaneously absorbed.

This change occurs at every inspiration, and is accompanied by an alteration of colour—the corpuscles, from having been before of a bluish purple, now assume the florid hue, and pass away to the left side of the heart to make way for the venous impure fluid behind them. So you see, when the breath is drawn in the interchange begins, the pure air is absorbed and carried away by the blood into the system, and the impure fluid, the vapour, and some organic matter are expelled in the expiration. To prove that these materials are emitted by the breath, perform the following experiment. Take two bottles, connected as shewn in the diagram (fig. 45), by corks and tubes; place in one (*a*) some clear oil of vitriol, and in the other (*b*) a small quantity of lime-water, and then expel the air from your

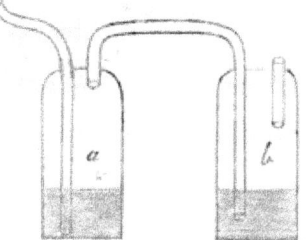

Fig. 45.

lungs through them, for about a dozen expirations, and you will observe that, first, moisture has collected on the sides of *a*; second, the oil of vitriol has become blackened; and third, the lime-water, which was before quite transparent, presents a milky appearance, which three phenomena demonstrate, that the breath contains water, organic matter, and carbonic acid; for organic matter blackens oil of vitriol, and carbonic acid whitens lime-water. It was formerly supposed that the carbonic acid which emanated from the lungs was formed in these organs, that the venous blood contained a large quantity of charcoal matter, or carbon, which combined directly with the oxygen of the air—in fact, that combustion took place in the lungs themselves. This view, when I come to speak of the cause of animal heat, we shall perceive to be incorrect, but, in the meantime, on the supposition that it is not erroneous, let us try an experiment to test its truth. If the combination originates in the lungs, it does so because the carbon of the blood unites with the oxygen of the air; thus

$$\left. \begin{array}{l} \text{Carbon} \\ \text{Oxygen} \end{array} \right\} \text{Carbonic acid.}$$

If, then, we cause an animal to respire a gas which does not contain a trace of oxygen, it is evident that carbonic acid cannot be generated.

Experiment: Place a bird in a glass jar filled with pure hydrogen gas, in which it will live for a short time, and after it has remained for a certain period remove it, and *carbonic acid will be found in the vessel*. Therefore, we see that it is not formed in the lungs, but in different parts of the body, and on its arrival at the pulmonary organs is simply discharged from the blood—oxygen taking its place.

Up to this we have been regarding the air as composed of oxygen alone, and we have been doing so for the sake of simplicity; that it is really a mixture of two gases I dare say most of my readers are aware.

$$\text{The atmosphere contains in 100 parts} \begin{cases} 23 \text{ oxygen,} \\ 77 \text{ nitrogen,} \end{cases}$$

with just a vestige of carbonic acid, and a variable quantity of oxygen in disguise, and with concentrated power, known as *ozone*. The properties of these two chief constituents are exactly opposite; thus, oxygen supports life and flame; nitrogen extinguishes both. The first is the great vivifier, and if breathed in its purity would cause the vital processes to be carried on with too great a velocity; the second possesses negative qualities, and its use is to dilute the oxygen, in order that the animal and vegetal functions may be performed *steadily*.

The relative actions of the two gases are well exhibited by igniting two pieces of phosphorus, and dipping one into a jar of oxygen and the other into a vessel of nitrogen; the first will burn with such brilliancy that the eye can with difficulty gaze upon it, the second will be instantly extinguished. The two fluids are only *mingled* in the atmosphere, not chemically combined, and an artificial mixture of the two will form a gas indistinguishable from ordinary air. It is sufficient to excite our astonishment to know that in the course of a single year we consume 100,000 cubic feet of air, and purify about 3,500 *tons* of blood. An animal may be suffocated from absolute want of oxygen in the atmosphere; but even when this gas is abundant, the presence of a certain proportion of carbonic acid may produce smothering also, as we know from the reports of the many cases of charcoal poisoning which have occurred in France. If a sparrow be placed under a bell-glass, it is found that after a certain period he dies of suffocation; if, however, there be placed also beneath the jar some substance, such as potash, which will absorb the carbonic acid, life is preserved for a much greater length of time.

How it is that the presence of foul air operates, is not well understood. Some have attempted to show that it prevents diffusion, but the arguments in support of this idea are anything but satisfactory, and it is far more likely that it exerts a peculiar poisonous action on the lungs, in this way depressing their power of performing the proper function. That it gradually lowers the various animal processes, is evident from the effect produced on persons who have been respiring a vitiated atmosphere. For example, in an overcrowded church, or theatre, or concert hall, who has not

"most somniferously sleepy felt?"

A curious fact in connection with *this* sensation of drowsiness is that you gradually become habituated to the locality, your instincts of suspicion regarding the danger of your position are lulled to rest, and death is steadily creeping over your frame, whilst you are quite unconscious of his approach. Yet, if a second person enters the room whose atmosphere you have been tranquilly and unsuspectingly respiring, he is instantly warned of the danger, a sense of suffocation oppresses him, and he rapidly withdraws. *Your* lungs, by the gradual and uniform depression of their activity, have become inured to the conditions which surround you, while *his*, in the midst of their activity, have received a sudden and severe shock. An ingenious experiment of a great French physiologist demonstrates this more fully. A sparrow is placed, as in the last instance, beneath a bell glass, and retained there till life begins to wane; he is then removed, and rapidly recovers; if, however, another, a vigorous and healthy bird, be introduced into the atmosphere vitiated by the first, he almost *instantly expires.*

The importance of attention to ventilation cannot be over-rated, and we do not require to be reminded of such sad occurrences as those in the "Black Hole of Calcutta," to appreciate the advantages resulting from a suitable supply of fresh pure air. It is to be lamented that hygiene is so much neglected as a science in this country; and especially that hospitals, asylums, and workhouses should be erected on sites chosen (as often happens) by individuals who, either through ignorance or prejudice, are unfitted to make a proper selection. It is melancholy to reflect upon the fearful consequences of inattention to the sanitary condition of the "lane and alley" portions of our large towns. Typhus, cholera, scarlatina, and such like, are in many instances the offspring of filth; and that filth in its most loathsome and obscene forms exists in the purlieus of large towns, I think few will deny. The larger streets, through which my lord's carriage rolls, are paved and drained, and swept and watered; but the narrow entries, the obscure side streets where the poor reside, and from which the stench proceeding is often intolerable to all but those inured to it—who cares for these? The corporation? No; your big-bellied alderman, and bloated mayor, cannot be seen in such disreputable localities! Their duties relate to the public—but then those *common people!* you know it is utterly impossible to teach them cleanliness! Oh, of course, of course, Mr. Mayor; but might I ask you when did you make the attempt? If it be deemed advisable that borough analysts should be appointed to prevent the imposition of adulterated substances upon the public, which, when practised even extensively, injures even at the worst but a few, how much more it is to be desired that some fit person be employed to see that as little adulteration as possible of the air we breathe—of the food of all, from the king to the beggar—be admitted! Dear reader, if you have ever experienced the atmosphere which prevails in the dens of the poor—if you have ever toiled, flight after flight, over the rickety, rotten staircases of obscure lodging-houses, and entered the gloomy, ill-lighted, ill-ventilated apartment of some dying patient, and observed parents and children huddled together in sickness, squalidness, and misery—you will forgive me this digression.

The way in which we breathe has, I trust, been already made sufficiently apparent; the reason why we do so is not quite so evident; in fact, if truth must be told, is not known at all. I do not mean to say that it has not been explained; for I believe there is hardly a work on physiology extant but what contains a lengthy explanation (?) of the cause.

Don't for a moment suppose that I would accuse the philosophic authors of endeavouring to deceive the public; but this is pretty certain, if they did not intend to confuse their readers, they assuredly have obfuscated themselves. A recent book-maker seems to have discovered the entire case, for he writes in the most unhesitating manner,—"The respiratory acts are automatic, occurring without our being aware of them;" of course, the question is a delicate one, but since the gentleman is not aware of the existence of his own respiratory movements, I should like to inquire how he obtained the information. Further on he observes of the respirations, that they are "produced by the unærated blood in the lungs," giving rising to a "besoin de respirer." This is cloaking one's ignorance, with a vengeance. The unærated blood producing a *necessity* for breathing is the *reason* why we *do* breathe. I need not remind the reader of a certain animal which once disguised itself in the skin of a lion, and through its peculiar vocal intonation confessed its real character; but it does strike me that the above philosophic generalization is a suggestive one.

External impressions often bring about the action of the lungs when it has apparently ceased; thus, water thrown upon the face or chest of one who has just fainted, rouses the dormant power and restores consciousness of surrounding objects. Similarly in the newly-born infant the function of respiration is stimulated to performance by the influence of the cold air upon the face, or artificially, by a not over gentle thwack imparted to the more ignoble extreme by the hand of the attendant "Sairey Gamp," if so to do "disposed." Of the way in which this external stimulus acts, or the road by which the effect is, so to speak, conveyed to the lungs, we know literally nothing.

As bearing on the subject of respiration, I think we may, with advantage, here consider the kindred one of *voice*. I must, however, premise my remarks by telling you that the country we are about to explore is not a very level one, but with hill and dale, upland and low, mountain and moor; so that though we travel through the different districts, the contour of the surface is so varied—I might almost say so complex—that at the end of the journey you will not acquire so much information as possibly you anticipate.

Voice is a general term, and includes under it two forms:—first, voice proper, or mere musical sounds; second, the different classes of articulation or speech. And we find that voice and speech, although both due to the expulsion of air from the windpipe, yet are produced by two distinct sets of apparatus; the former owing its production to the larynx, the latter to the mouth and tongue. Two experiments demonstrate the truth of this statement. If an opening be made into the windpipe, below the larynx, voice is lost so long as the incision allows the air to pass through it. And to prove that the utterance of words is owing to the mouth and tongue, a tube has been introduced through the nostril* so as to reach the back of the mouth, then the breath having been held, air has been blown through this tube, and, by the evolutions of tongue and lips, sentences have been uttered in a faint whisper.

The larynx is an apparatus composed of an intricate assemblage of muscles, gristly rings, and membrane. It stands on the upper end of the windpipe—to which it is joined—as a statue rests on its pedestal; and above it is covered by the root of the tongue. Its largest cartilage (or gristle), the *thyroid* † (*b b* fig. 46), is one with which everybody is familiar under the name of *Adam's apple*, but which is not confined to man, being developed, though to a lesser extent, in woman also. This, at the top, forms the doorway of the organs of respiration. But this doorway is a "*trap*," that is, it does not stand vertically, so that the tenant can enter and depart in the erect posture, but is horizontal, so that he must go out head foremost. Its sides, or "jambs," are com-

---

* That a communication between the back of the mouth, or pharynx, and the nostril *does* exist, the reader, who, in an explosion of laughter at the breakfast table, has sent his coffee, viâ his nose, into his pocket-handkerchief, will at once admit.

† From two Greek words, signifying a "shield" and "like."

posed of two folds of membrane, meeting to a point in front, and separated behind, which are termed, improperly, vocal *chords* (*e e*, fig. 47). These, by means of peculiar contrivances, can be shortened and approximated, so as to alter, within certain limits, the size of the aperture, which is, as you will have concluded, of a triangular outline. So much for the doorway. It opens into the throat, or highway leading to the gullet, and, on that account, an eccentric or vagabond morsel of food

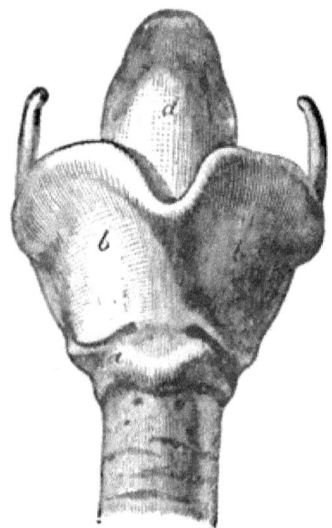

Fig. 46.

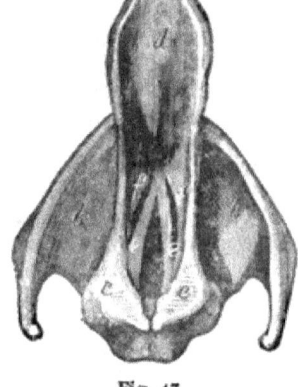

Fig. 47.

Fig. 46.—Human Larynx, seen from the front. Fig. 47.—Same seen from above. *a*, the ring-gristle; *b, b*, the thyroid; *c, c*, the pyramidal levers; *d*, the epiglottis; *e, e*, and *f, f*, the vocal chords.

often, regardless of its proper destination, might succeed in gaining admission to the lungs and doing a deal of mischief, were it not for the door which is closed against it. This door is called by anatomists the *epiglottis** (*d*, fig. 47). It is, in outline, like a heart-shaped leaf, is hinged at its point to the front of the ring known as Adam's apple, and laps when closed, over the doorway and its sides, the vocal chords, as shown above. This closure of the epiglottis takes place on every occasion of swallowing, and so food particles of a migratory turn are prevented passage into the windpipe and wend their way stomachward. The mechanism by which the "vocal chords are brought together and separated is ingenious and interesting. In front, as I mentioned, the two cords (*e e*, fig. 47) are attached to the "pomum Adami;" behind, however, where they divaricate, each is conjoined

* *Epi*, upon, and *glottis*, tongue; a derivation which affords a misconception as to the true position of the structure alluded to.

with a little gristly pyramid (*c c*, fig. 47), being united to one of the corners at the base, and since each pyramid moves on a central pivot, and various muscles are connected with the sides and other portions, it will be readily seen that by the movements of the pyramids in one direction or the other the chords will be separated or approximated.

The pyramid is a lever of the first-class, like that of an ordinary balance; when one end is pulled inwards, the other is moved out, and vice versâ; and as the vocal chords are attached to one extremity of each, they necessarily are caused to approach or diverge according to the direction in which the pyramids are drawn.

When it is required to tighten the cords, the gristle (*b b*, fig. 46) to which they are attached in front is drawn downwards by a special muscle. Thus we perceive that this orifice, guarded at its sides by the vocal chords, is capable, by an approach of these latter, of being diminished in size, and of being also made tense or slack as is needed.

Have you ever opened an accordion or concertina, and seen all the brass tongues of different sizes, by the vibration of which the various notes are produced? If so, you can at once catch the meaning of the human hautboy—the vocal chords—for these are not to be likened to a pair of strings, nor to the flute pipes of an organ as is occasionally done. They constitute really a sort of reed, which, by virtue of the appliances I have been describing, can be made large or small; and this shortening can take place to an almost inconceivably slight extent, as we shall see presently.

In man each chord measures seventy-three hundredths of an inch when relaxed, and when *most* tense ninety-three hundredths, or about one-fifth of an inch more; and, as every variety of sound is produced by a shortening or lengthening of the chords within the limits of that one-fifth, how immeasurably minute must be the difference in length necessary to the production of two distinct sounds in the same octave? The longer the chords the lower, as regards the musical

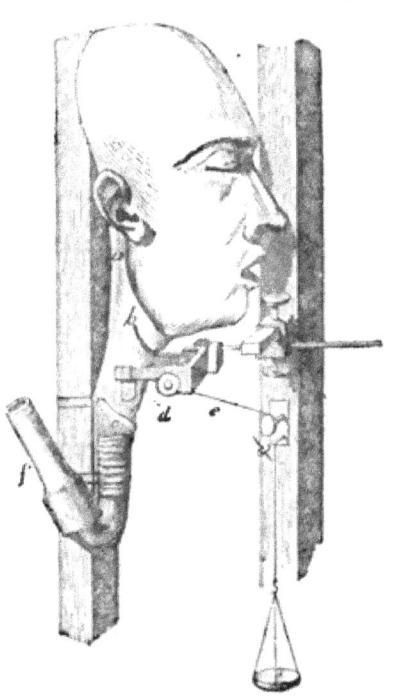

Fig. 48 represents the head of a corpse arranged for experimentation on the organs of voice. *a b*, the throat; *c*, the top of windpipe; *d*, an apparatus by means of which the orifice of the larynx can be altered in size; *e*, a string which passes over a pulley and has attached to it a scale pan; by placing a weight in the latter, the string is made tense, and the Adam's apple brought forward; *f*, a tube connected with a pair of bellows.

scale, will be the sounds produced by them; and, therefore, as a man's voice is more of the bass kind than a woman's, we find his vocal chords to be the longer of the two. In the female they measure, when most drawn out, only sixty-three hundredths of an inch. Here a fact suggests itself: a boy of twelve has a voice resembling a woman's, but at the age of fifteen, or thereabouts, it "cracks," and becomes coarser. Why is this? Because at the same period in which the voice alters its character, the Adam's apple becomes enlarged, and is seen to project further, and it is to the inner surface of these the chords are attached; so that, as they follow the advance of the gristle, they increase in length, and hence the voice is changed.

The compass of the voice of an amateur singer is, at the lowest, about two octaves, each of which is equivalent to about twelve semitones, and since it is possible easily to produce ten distinct musical sounds within each semitone, the number of distinct intonations which may be produced by the organs of voice can fairly be set down as two hundred and forty. Now these are, as you know, produced by different degrees of shortening or lengthening on the part of the chord, and, as the limit of increase or decrease is one-fifth of an inch, it is clear that in passing from one sound to that immediately above or below it, the chords are diminished in length or elongated by the two-hundred-and-fortieth of one-fifth of an inch, or, in other words, the *one-twelve-hundredth* of an inch. From what is known of the vocal powers of the distinguished Madame Mara, it has been calculated that her voice-chords can only have been altered in length, by about the *ten-thousandth of an inch* in the production of each separate sound. These facts are, *indeed*, sufficient to arrest our thoughts, and should teach us to reflect upon the wondrous foresight and intelligence of the Almighty Jehovah. We are told with astonishment of the extraordinary adapting power of Nasmyth's steam-hammer, which can crush an anchor or drive a nail with equal facility; but what is it in comparison with this little mechanism—the organ of voice—in which two simple folds of membrane can be tensed or relaxed to the ten-thousandth of an inch with the most mathematical exactitude, and which, in those that have the mighty gift of song, can send forth—

"Such floods of delirious music,
That the whole air, and the woods, and the waves seem silent to listen."

Some persons are remarkable for the loudness of their voices, others for the softness, and though these qualities are said to depend upon the force with which the air is driven from the lungs, it seems more probable that they are due, as it were, to the echoing power of the larynx itself; for in monkeys called Mycetes, which inhabit America, there are hollow sacs attached to the gristles of the vocal organs, and the effect of which is to render the howling of these creatures a source of great terror to travellers, it being asserted that their nocturnal concerts may be heard for miles, the sounds being even more powerful than the roaring of Herr Formes, or the lion itself. The cause of variation in the "timbre" of the voice appears to be a difference in the amount of hardness or softness of the organs, but of this we know *nothing* with certainty.

This is the place to *say a few words about speech*. Sentences are combinations of words, these latter of syllables, and syllables of sounds. Now how do we make those different sounds? The vowels are formed simply by the increase or decrease in size of the mouth, or by an alteration of its shape. The consonants are formed by various movements of the tongue and lips, as the reader may see for himself by "running through" the alphabet, and marking the process by which each sound is produced. That speech results from movements of the tongue and lips has been demonstrated ingeniously by the invention of speaking machines, which have been exhibited from time to time, one of which was so perfect that it could utter many distinct sentences, and among others, "I love you with all my heart."

Stammering is caused by a want of control over the muscles employed in the pronunciation of certain letters, and is, to all intents and purposes, similar in its origin to the disease called St. Vitus's dance, in which it is found impossible to govern the movements of the muscles of the limbs. Of this we may be assured, that the only cure for the stutterer is caution in speaking and reading. Let him, if possible, feel assured that he *has* the power, but let him exert it slowly and with care. Ventriloquism is a faculty, the nature of which every one is conversant with; but when we come to inquire how the peculiarly deceptive voice is produced, we are somewhat in a difficulty. One great physiologist prides himself, doubtless, upon the discovery that ventriloquism is only the imitation of sounds produced at a distance; but I fancy many of my readers have made the same discovery, which is, at best, but an expression of the deceptive powers exhibited by professors of the art. The derivation[*] of the word would imply that speaking in the stomach was the power of those gifted with the ventriloquistic faculty; but a mere tyro in anatomy would perceive at once the utter absurdity of such an opinion. The German view is certainly the most probable one. It is this: the sounds of the ventriloquist are produced by taking a deep inspiration, in doing so, depressing the diaphragm and protruding the abdomen, and then breathing out slowly by compressing the lungs with the lateral portions of the chest—the sounds being modified by the mouth as in the case of ordinary speech.

Whistling is the production of musical sounds by forcing the air rapidly through a small orifice formed by the two lips. It is due to the series of vibrations into which the current of air is thrown in passing from the mouth; because, if air be driven with rapidity through the mouthpiece of a flute, a similar series of sounds is developed by altering the aperture.

Sighing is caused by any circumstance which tends to depress the system, and is a form of deep inspiration, which originates in a want

---

[*] From *venter*, stomach, and *loquor*, I speak.

of air in the blood. Thus, if a person be very much absorbed by any subject, the nervous power is, so to speak, drawn from the pulmonary organs into some other channel, and afterwards, to compensate for the feeble exertion of the lungs, a series of deep inspirations follows— the individual sighs. So that, as Mr. Lewes truly observes, "the philosopher brooding over his problem will be heard sighing from time almost as deeply as the maiden brooding over her forlorn condition."

In yawning a deeper inspiration is taken than in sighing; but the cause is pretty much the same. Why it is so catching is quite an enigma; and much as men may talk about sympathies, and such like, it does not appear that we are able to solve the problem.

Sobbing and hiccup are expiratory, or breathing-out actions, when the door of the larynx is partially closed. Coughing is what is termed a reflex* action. It is produced by some substance— mucus, cold air, food, &c.—irritating the larynx, windpipe, or bronchial tubes. The irritation stimulates the muscles of the chest and the diaphragm, which then contract violently in order to expel the intruder. It is called reflex from a peculiar view as to the way in which the effect of the irritant is transmitted. Thus in the spinal chord lies the power of bringing into play the muscles employed in respiration, and it calls this power into operation by means of nerves passing from it to the parts alluded to. But it also has the power of receiving impressions through nerves which travel *to* it from the lungs, larynx, &c. Now, in a reflex action, the irritant (a morsel of food which has lost its way) produces an effect upon the nerve; this effect is conveyed to the spinal cord, which then transmits the power to the muscles, that they may expel the stranger—the impression being as it were *reflected*.

Sneezing is a more intense form of expiration than coughing, and is similarly caused, the irritant particle being in this case situate in some portion of the nostril. That it may arise from the irritation of other parts is also certain—a distinguished German physiologist being obliged to sneeze whenever the bright light of the sun falls upon his eyes. The use of the pocket-handkerchief, and indulgence in a hearty blow of the nose, is often sufficient to terminate a fit of sneezing; and I can particularly commend the practice to those elderly gentlemen who are the terrors of the tea-table when they commence, after the fashion of "Mr Staggers," looking alternately at their friends and handkerchiefs.

The way in which laughter is produced has not been clearly made out; although, doubtless, this very statement, from its apparent untruthfulness, may excite a smile. Granted, we know how to make a friend laugh; we may tickle his ribs with our fingers, or his fancy with a joke; but we cannot tell why it is that he laughs, after all. The most plausible explanation is this: a good joke gives rise at once to an idea, and this to a certain quantity of nervous force, more or less, according to the temperament of the recipient. This nervous force, having been developed, must be expended, and the

* The objections to this view would not be adapted to appear in so popular a treatise as this.

expenditure may take place in one of two modes:—firstly, in developing a series of new ideas till it is exhausted—this is the effect upon a thoughtful man of a nervous temperament; secondly, by being transmitted to the muscles, and calling them into play as in laughing. But you will say, if I relate a funny anecdote to a man who does not laugh is he therefore thoughtful? No, he may be a man of lethargic turn, and your tale has developed no nervous force at all. Why does not the nervous force, when carried to the muscles, affect those of the arm and leg as well as those of the face and chest? This is a very fair query, and I admit is the one most difficult to answer. Possibly because it flows into the nerves of the face and respiratory organs more readily than it could pass into the lower nerves, and so affect the limbs; and it is familiar to all of us that when there is an exuberance of the jocular laughing force, it *does* affect the limbs also, as exhibited in that tendency to throw up the arms and legs when—

"O'er all the ills of life victorious."

The laughter which results from physical tickling would come under the category of reflex actions; and as we cannot explain it, we shall leave it, along with many other vital phenomena enjoying that physiological limbo.

Ere we conclude this chapter on respiration, let me caution you against confounding two very distinct diseases of the lungs: consumption or phthisis, and bronchitis. The former is a malady affecting the fine tissue of the organ—air cells, blood-vessels, &c. The latter is an inflammatory state of the bronchial tubes and windpipe. Below, I have given in a tabular form a few of the characters of each; but above all things let me impress upon you the necessity of consulting a physician at the very outset of either disease, for of "all the ills that flesh is heir to" there are none in which a really scientific practitioner can effect more for the patient than in consumption and bronchitis, if the case be undertaken at an early period of the attack; and none which are so likely to be confounded by the unskilled. An anxious mother may think her daughter consumptive because she is subject to catarrh, or may imagine that a really insidious attack of consumption is but "*A mere cold.*"

In Consumption.	*In Bronchitis.*
1. Shiverings and flushings.	1. No shiverings or flushings.
2. Cough comes on in middle of night, or toward morning.	2. Cough on rising in the morning and on going to bed at night.
3. Pulse is very rapid.	3. Pulse beats as usual.
4. Disease of the stomach and loss of appetite; and in females general derangement of system.	4. No disease of stomach.
5. Inability to eat fats and sugar.	5. No distaste for fats or sugar.
6. Pain in upper part of chest, beneath *collar* bone.	6. Pain of itching character, just beneath *breast* bone.
7. Very often diarrhœa.	7. No diarrhœa.

CHAPTER IX.

Heat—Centigrade and Fahrenheit's Thermometer—Why we feel Cold—Normal Temperature of the Body—Theory of Combustion in the Lungs—Doctrine of Liebig—Heat without Flame—Heat-forming Foods—Sources of Animal Heat—Influence of the Sympathetic Nerve—Tolerance of intense Heat by the lower Animals—The "Fire King."

Man is a warm-blooded animal. I fancy I hear the reader say to his less learned friend, What is the meaning of the expression warm-blooded? Having the blood warm? That, my dear friend, is no answer to my question; for you are still unable to tell me what being warm *is*. What you meant when you said warm-blooded, was this: the circulating fluid is of a greater temperature than that of the surrounding air. I thrust the poker in the fire till it is red hot, and then lay it aside, and what do I find? The fiery metal gradually becomes black, and its heat passes away to the air, and to everything in its neighbourhood, till at last it is quite cold again. Is this the case with man? You will reply, Yes; sometimes he is hot, at other times he is cold. Wrong; his blood is always at the same heat. If a thermometer be placed under your tongue, the mercury will rise in the tube till it stands at 98 deg., but it will go *no higher;* nay, if you were perched on one of the glaciers of the Alps, or were being broiled beneath the sun of Hindostan, it would just indicate the same degree—98. Of course I refer to Fahrenheit's thermometer, which is the one usually employed in this country, and not to that which is termed the "*centigrade.*" In our thermometer the mercury rises to the number 212 when placed in boiling water; in the centigrade, however, the boiling point is indicated by the number 100; so that you must bear in mind that the blood-heat degree, 98, of the thermometer employed in England is almost the boiling one of the French instrument.

Some years since a gentleman was recommended by his physician to travel to the south of France for the benefit of his health. He followed this advice, and early in the spring set out upon his journey to the southern provinces of that country. Shortly after his arrival, he determined to indulge in the luxury of a warm bath, and on being asked by the bath superintendent at what temperature he required the water, replied at about 100 deg. The manager mildly expostulated with him, but in vain; and I suppose, having had some considerable experience of the determination characteristic of the Anglo-Saxon, at length yielded to his entreaty, and prepared the bath—muttering to himself as he went out that the poor gentleman would certainly be scalded. The Englishman, however, was not devoid of caution, and being led to suspect, from the opposition of the superintendent and from the excessive quantity of vapour, that something must be amiss, quietly dipped in his great toe to test the temperature, and finding to his dismay that he had been on the verge of giving himself a lobster's death, left the establishment much discomposed, and on arriving at his hotel was informed of his mistake; which mistake was not a very extraordinary one, but such as the reader must recollect when desiring to enjoy a warm bath in France.

You no doubt consider it very strange that you are not cold when you feel cold, nor warm when you experience a tendency to adopt Sidney Smith's recipe—" take off your flesh, and sit in your bones." Yet such is the truth. When you shiver, you are really, as regards your blood, no colder than when you felt quite comfortable as you sat before your fire. Sensations of heat and cold are of course common; but even whilst we experience them, our blood is not colder nor warmer than usual.

We say we are warm, because the blood circulates more rapidly in the superficial portions of the body—as the skin, and then an effect is produced upon the nerves which supply those parts, and a corresponding sensation developed. This is very nicely shown in ague. Here there are shivering and perspiring fits alternating with each other. In both the temperature of the blood is proved to be the same by the application of the thermometer; but when the patient complains of the intense cold, and the trembling movements characteristic of the disease come on, then it is found that the blood has left the surface, and the internal organs are overloaded with it; whilst on the other hand, when the flushing and perspiring attack makes its appearance, the vital fluid has returned in unusual proportion to the skin, and has left the viscera. This demonstrates conclusively that the sensations of heat and cold do not prove that *actually* the temperature of the body has been altered. Therefore, it is quite clear that few persons know what heat *is*; and many will be surprised to learn that what they understand by the term is little more than a name which is given to one of the effects of heat. Heat is really —— but, stay! What is this? "All persons trespassing on these grounds will be prosecuted." Why, dear me! Where are we? Oh, I see! Professor Faraday's domain. Let us retrace our steps as quickly as possible; for if we ventured any further it would not be to our advantage.

I was saying that man's blood is always maintained during life at a temperature of 98 deg.; the question then arises, how is this effected? Heat is derived from two sources*; the sun, and chemical combination. The second you may not at first recognize; *it* is the mode in which heat is produced by an ordinary fire. For example: the carbon of the coal unites with the oxygen of the air, and forms heat and flame, the smoke being composed of minute particles of the carbon which have not been quite burnt, and which therefore are so much valuable material expended uselessly.

---

* It is possible that heat of all kinds is only physical motion of minute particles of matter, even that the heat of the sun is due to the effect of its attraction of the atoms of atmospheric air; but the old view is the more intelligible to ordinary readers.

Can we now apply this theory of the production of heat to explain the process in the human body? Yes, in some measure; but we must not imagine that all the heat of our blood results from *any* kind of combustion. Many years ago a great French chemist put forward the view that the temperature of the body was due to combustion; in fact, that the carbon of the blood joined the oxygen of the air, and developed heat, this phenomenon taking place in the lungs. I here anticipate a question you are about to ask, by telling you that combustion may take place, and heat may be formed, *without flame*. The manure in a dung-pit is very warm, the heap of leaves in the adjoining grove, which the cruel winds of autumn have severed from their parent branchlets, and which we turn over with our walking-stick as we pass along, are reeking hot. See how yonder hay-rick steams! Thrust your hand into its side, and feel the scalding temperature of its interior, and then say there cannot be heat without flame. In all the instances I have alluded to, combustion has been taking place, and the carbon, whether in the dung-hill, the leaf heap, or the hay-rick, is slowly but certainly uniting with the atmospheric oxygen.

So you see there was some reason in this doctrine, that the lungs were the furnace of the body; and popular lecturers were wont to say that the ribs were the bars of the grate, and the diaphragm the huge forge-bellows, which propelled the air through the burning fuel. This was all very nice, and exceedingly simple; but, alas! like all *exceedingly* * nice theories was destined to succumb to science. Unhappily for this idea, it was found—firstly, that the furnace was no hotter than the water tubes which carried the heat—that the temperatures of the lungs and the little toe were identical; and secondly, that the quantity of carbonic acid generated was, occasionally, *much greater* than could have been formed from the amount of oxygen taken in by the lungs. Advancing science tolled the curfew, and out went the lung-fire theory, never to be rekindled.

From its ashes there sprang, Phœnix-like, a second hypothesis—that of Liebig. This philosopher contended that, though combustion did not take place in the lungs, yet that certain alimentary materials were *really* burnt in the capillaries distributed throughout the body. In carrying out his opinion he divided all sorts of food into two classes.

Food
- Heat producers { Fats, oils, butter. / Starch, sugar, &c.
- Flesh producers { Meat, eggs. / Gluten, &c.

On the supposition that all the aliments of the second group went to build up the worn-out tissues, and all the others to maintain the animal heat.

* I need not here call attention to the contest now taking place between the advocates of creation-of-species-ism, and those who support the grand generalization of Mr. Darwin.

SOURCES OF ANIMAL HEAT.

According to this view, we should regard the microscopic blood-vessels of the whole body as collectively constituting a sort of oil or spirit lamp, in which are burnt the fatty and alcoholic portions of the food, combining here with the oxygen introduced into the arterial blood through the medium of the lungs. It is a hypothesis in part correct, and, to some extent, erroneous; for, as I observed in an earlier chapter, we must not think that the only office of the oily and starchy portions of the food is to keep up heat, because they also enter into the composition of the tissues; and furthermore, we *know* that heat is produced apart from the combustion of food. When a mixture is made of oil of vitriol and water, a very intense heat is formed, and the same thing holds good for many other mixtures. When salts of various descriptions are decomposed, or caused to unite with each other, the temperature is increased; and since we know that hundreds of such changes are being made in the animal body during every moment of our lives, we have here sufficient proof that food is not the only agent in the production of warmth. Again, it has been shown that when animals were fed upon food which contains neither fats nor starch, the heat of the frame was, nevertheless, kept up. Plants which, in the daylight, do not cause carbon and oxygen to unite, but even separate them from one another, possess heat to a certain extent; whilst, during the night, when their carbon combines with the oxygen, forming carbonic acid, as in the case of man, the quantity of heat generated is *much less* than that formed under the influence of the sun; and for this reason, during the day, the different processes of repair are going on with rapidity; but during the night they are almost quiescent. A favourite argument of those who would have us put our trust in Liebig's doctrine, is one deduced from certain experiments on pigeons. These birds were kept fasting, and it was remarked that the temperature fell very much; hence the conclusion that food is the generator of heat. This, however, is only special pleading, for after some days of starvation, the temperature of the pigeons *rose* again; thus, more effectually than ever, putting the extinguisher on Liebig's supposition that fats, starch, and such like, are the sole source of animal heat.

Friction gives rise to increase of temperature. Rub two pieces of board together, and mark the heat produced; or pull a cord rapidly backwards and forwards over a rough surface, and observe how hot it will become. We can fairly presume that warmth is developed in the blood-vessels in a somewhat similar way; firstly, by the velocity with which the stream of fluid is driven through the vessels; and secondly, by the friction or rubbing of the blood disks against each other. In summing up, we may assert that the heat of the body is derived from three distinct sources:—

1st.—The combustion of certain portions of the food.
2nd.—The combination and decomposition of different tissue materials.
3rd.—The friction { *a* Between the blood and sides of vessels.
{ *b* Between the blood disks themselves.

Some parts of the nervous system appear to exert great control over the temperature of the body. These portions are, collectively, called the sympathetic nervous system; and such is its influence that if the branch of either side be divided, it is found that the regions supplied with filaments from it have their temperature greatly increased. This alteration is thus accounted for: the capillaries are provided with two sets of nerves, whose actions are antagonistic; the sympathetic nerve, by acting on the coats of the vessels, tends to lessen their calibre, and the ordinary nerves operate in such a way as to cause an increase of this latter. When the sympathetic is cut its influence ceases, and then the ordinary nerves, by virtue of their opposite power, cause the canals to expand. When the vessel is enlarged, it contains a greater quantity of blood than usual; hence there is a greater amount of chemical change and more friction, and, therefore, an increase of temperature, which increase often reaches to as much as eleven degrees above the ordinary heat.

It is exceedingly interesting to mark the difference between the higher and lower animals regarding the effects upon them of diminution of temperature. For example, while we can freeze the grubs of certain moths without injury or destruction to life, if we lower the temperature of man's blood to sixty-eight degrees he ceases to exist. In one instance, the larva of an insect had been so completely frozen that when thrown into a glass vessel it "chinked," as a stone would, yet, when heat was applied, life was restored, and after the proper interval the maggot became a perfect moth.

If the blood of man be elevated about thirteen degrees above the natural temperature, life is extinguished, although you can virtually *boil* some of the lower creatures (wheel-animalcules) without destroying vitality. You may possibly ask, If this be the case, how can a person enter an oven, heated to 500 or 600 degrees, without sustaining any injury?—a feat which has been over and over achieved by "The Fire King," Chabert, and others. The reply to this I must defer to the next chapter, as it involves a reference to the organs situate in the skin, and to the effects of heat on water, of which we have, up to this, said nothing.

CHAPTER X.

The Skin—Epidermis or Cuticle, and Derma—Corns—Structure of the Papillæ—Perspiration—Form and number of Sweat-glands—Twenty-seven miles of Perspiration Tubing—Composition of the Perspiratory Fluid—End attained by the action of the Sweat-glands—Injurious effect of Lactic Acid when retained in the system—Experiments on Dogs—How the temperature of the Blood does not exceed 98° Faht.—Why a man can enter with impunity an Oven heated to 600°—Impropriety of employing the expression Latent Heat—Transportation of Force when Solids pass into Liquids, and the latter into Gases.

THE skin is the tissue, or membrane, which clothes the surface of the body externally, and is composed of two layers, an inner and an outer, separated by a thin bed of gelatinous substance. Let us take a thin slice of it, and place it beneath the microscope! We perceive that one side of the specimen is composed of a vast number of interlacing cords, or bands, which end in a dark space, and on the other side of this we see several rows of particles, the first of which are somewhat oval, but these as we advance become spindled-shaped, and the outer ones are quite scaly—three or four being united, and flattened out. We also observe that the intervening dark line is thrown into a number of wave-like folds, to which the first row of particles conforms. The outer stratum is called the *epidermis*, or cuticle (*a*), the inner the *derma* (*b*), and both grow from the dark line,—the oldest part of the first being the outer, and of the second the inner surface (fig. 49). There are no blood-vessels in the epidermis, and this is the reason why a person may pare a corn, or cut an old blister, without bleeding from the parts so operated on. Its only object is to protect the important tissue beneath, and it has a peculiar power of growing fastest where most pressed on, as if nature foresaw the necessity and made the provision. Therefore we find the hands and feet protected by very thick layers of cuticle, whilst on other parts, as the lips and cheeks, it is extremely delicate. But, as any good quality may by a combination of circumstances be converted into an evil one, so is this power of the outer skin to grow quickest where there is most pressure, in many instances an exceedingly unsatisfactory one. Can

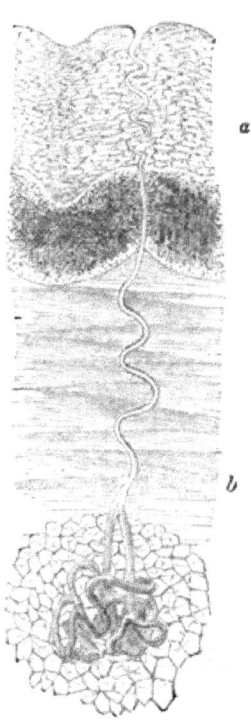

Fig. 49.

you ask why? Because, were it not for this property we should never be plagued with corns,—kinds of growth due entirely to the development of the epidermis, which is set a thickening by the squeezing of a cruel boot.

Beneath the skin lie the myriads of microscopic blood-vessels that give the body the peculiar colour which is absent after death. Along with them are placed the branches of the nerves, which endow the surface with the capability of receiving impressions. These nerves proceed outward as far as the derma, but go no further; and, as we shall see presently, they in many instances terminate in little elliptical bodies. Mark the hundreds of small furrows upon the under parts of your fingers. These, no doubt, have been often observed by you with the aid of a magnifying lens; but do you know what they are for? No. Well, let us first see how they are produced and then consider their purpose. You saw in the section placed beneath the microscope, a series of elevations alternating with depressions. The projecting hillocks are termed *papillæ*, and these in great numbers make up the lines upon your fingers, in which lines the sense of touch resides.

In each papilla is found a very minute, pea-shaped, solid body, round which a nerve passes, or into which it enters. When the finger is pressed against anything this little organ is displaced to some extent, and the effect produced on the nerve is then conveyed to the brain, giving rise to an idea or conception. Of these, more in another chapter.

Embedded in the deeper layers of the skin are many organs of minute size, called the *sweat-glands*. These are of two kinds: the first secrete a sort of waxy fluid, and are found well developed in the tube of the ear; the second are the most important, for, when taken together, they form an organ of a much more influential character than the kidneys themselves. It is a matter of no great difficulty to prepare a portion of skin containing one of these glands, and therefore you will suppose when looking at the diagram (fig. 49), that you are peeping through the microscope at the section I have prepared. Each of these organs consists of a tube, whose only opening is external, and which traverses the entire skin from its deepest to its most superficial portion. This tube or canal is twisted into a knot below (fig. 50), and lies in a little sac, which is surrounded by blood-vessels. For the rest of its course it wends its way in slight curves till it reaches the surface of the body, where its opening is situate. The sweat, or perspiration, which it is the office of these glands to form, is drawn into their knotted ends from the adjacent blood-vessels, and travels along the tubes till thrown out upon the skin. The number of sweat-glands is astounding, and ought to afford us some estimate of the necessity for attending to the state of

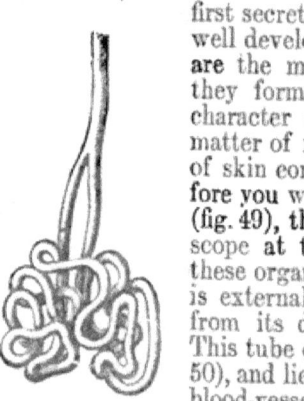

Fig. 50.

THE PERSPIRATION.

the skin. It has been calculated by an English anatomist, that in every square inch of the skin there are *at least* 2800 of these glands, and as there are in a person of ordinary size at least 2500 square inches of surface, it follows that the entire number of sweat-glands is *seven millions*.

Now, every gland tube when straightened measures a quarter of an inch in length, and this divided into the total number of glands will give (supposing the tubes arranged end to end) a length of tubing of 1,750,000 inches, or 145,833 feet, or 48,611 yards—or more than *twenty-seven miles*,—a system of sewage, beside which that of some of our cities sinks into insignificance. The perspiration fluid is of a sour character, and contains a certain proportion of lactic acid,* some organic matter, and various salts. It is constantly secreted, although we only perceive it when in large quantities,—as after much exercise. The average daily amount formed is one pound six ounces, but as it depends upon the extent to which the muscles have been exerted, it is hard to fix the exact quantity. By means of the perspiration, three ends are attained—

    1st. The blood is concentrated.
    2nd. An acid is got rid of.
    3rd. The body is kept cool.

Of course you see the way in which the first is achieved. The water is prevented accumulating by being got rid of in the sweat. If it remained, the fluid would distend the capillaries, and at length force its way from them under the skin, producing general dropsy; and this is the reason why, in such a disease, physicians prescribe remedies which, by acting on the glands of the skin, cause them to abstract the watery element of the blood. This, then, in its turn absorbs the fluid, which had made its escape under the integument, and so the parts are restored to their natural condition.

It is of the greatest importance that this lactic acid, to which I have alluded, be expelled from the system, for when retained it causes that most troublesome of maladies—rheumatism. Therefore, it should be remembered that constant exercise is one of the preventives of this disease; the exertion causes a flow of perspiration, in which the acid is poured out. It might, at first sight, appear an objection to this view, that even farm labourers are sometimes attacked by "the rheumatics;" but so far from upsetting the doctrine, it really supports it, as thus: A man who works much (physically) develops a greater amount of lactic acid than one whose occupation is of a sedentary nature; therefore any check to its removal will exhibit itself more fully in the former than in the latter. This check is produced by an exposure to cold, which brings down the temperature of the surface, and so prevents the formation of the sweat. When lactic acid was injected into the veins of dogs and cats, these animals were almost immediately affected with rheumatism; and when a post-mortem examination was made, the internal

* The acid of sour milk.

marks were the same as those in men who had died of rheumatic disease—a fact which demonstrates very fully the cause of this horrible ailment.

The third function of the perspiration glands touches on one of the most interesting series of phenomena in the whole category of vital actions. It is familiar to every one that when much muscular exertion has been made, the body becomes excessively warm; there is an increased flow of liquid from the skin, and the breath is quickened, as well as the pulse. Why do all these changes take place? We shall consider them one by one. When the muscles are employed, as in any violent exertion, they undergo extensive waste; this involves as extensive repair, and as the reparative materials are derived from the blood, it (by the form of attraction to which I have directed attention elsewhere) is drawn more rapidly towards them than before, so that the circulation is accelerated; but its velocity is also increased by the pressure of the muscles on the veins which carry the fluid to the heart. In this way the beats of the pulse and heart are more numerous per minute. The blood being brought more swiftly to the lungs than hitherto, the chest is stimulated; the air is then drawn in, in order to purify the circulating fluid, and by allowing it to pass away, to make room for that behind. Therefore the inspirations and expirations are more frequent. Greater heat is produced, because of the increased chemical change going on in the muscles, which are being *momentarily* repaired, and deprived of refuse materials, thus involving a series of combinations and decomposition of certain animal principles, which, as I have shown, raise the previous temperature.

The next question is, How is it that perspiration ensues?—1st. Because the blood-vessels which bring the nutritive materials to the muscles are also those which send branches **around** the knotted ends of the glands; as the circulation of blood in the former canals is accelerated, so is it in the latter, and the sweat organs being supplied in a given time with a greater quantity of blood, will secrete a proportionate amount of perspiration liquid; and 2ndly, Because during muscular exertion more than the ordinary supply of lactic acid is generated; and as it is the province of these glandules to expel this substance from the system, they are thus urged to secrete more quickly than whilst the individual was at rest.

Now the above sketch brings us to the consideration of what we put off in the last chapter, viz.:—the reason why (although the temperature of the blood cannot be raised to 111°) a man can enter an oven heated to 500° or 600° with impunity.

When by the action of heat a solid is converted into a liquid, as ice into water; or a liquid is changed into a vapour, as water into steam, a certain quantity of the heat employed vanishes *apparently*.

If, for example, you apply the flame of a spirit-lamp to a vessel containing water, it will, after a while, boil, that is, it will reach a temperature of 212°; but no furnace, not even that of the great Jewish king, could enable you to raise it one degree higher, so long as the steam could pass away. You see, then, that when water passes into the state of vapour, heat is lost (?). We can hardly say it is *lost;*

for, if we bring the steam or vapour back again to the condition of a liquid, the heat is experienced. Natural philosophers have given the term latent (hidden) to this quantity of caloric, which was so disguised that they could not detect it. But this is no explanation. The true cause is this: heat is not a distinct force, but only a form of the one force which exists all through nature, and which, when water is being changed to steam, is metamorphosed, or converted into mechanical power. Water cannot expand except when vaporised; it is vaporised by heat, which in this way gives birth to mechanical power.

Pardon this digression. Suffice it to say, that when liquid passes into steam, heat is lost. Now water can pass into vapour at the temperature of the body, and in doing so robs the body of its heat. This, then, is the mode in which the temperature of the body is kept at its normal standard. When a man enters an almost fiery oven, perspiration is formed very copiously; this, by the temperature of the surrounding air, is vaporised as fast as it is secreted, and the surface of the skin is found nearly as cool as before. If, however, the air of the oven were so charged with moisture, that it could contain no more (and, therefore, that no conversion of the perspiration into steam could result), then the temperature of the skin would be *intolerable*, and the individual exposed to it would literally be scalded to death.

It is on this account also, that on a day when the atmosphere is overloaded with moisture, we can take little exercise without being covered with perspiration which has not been vaporised, and that we complain of oppressiveness, &c. For this reason, too, we find the habit of wearing waterproof clothing so unpleasant — the secreted fluid being retained warm upon the surface of the skin, and not converted into vapour as it should be. Pour into the palm of one hand a little oil, and into that of the other some ether, and observe the difference in the sensations produced. The oil will not be vaporised, and will not chill the hand; the ether, on the contrary, will disappear in a moment, leaving such a feeling of cold as might be produced by holding a piece of ice. I fancy you are now in a fair position to grasp the idea. What a beautiful chain of phenomena is that exhibited whilst we are taking our routine walk of an afternoon. Now that we have unravelled the coil, and examined the different links of cause and effect which compose it, how infinitely intelligent seems the foresight which provided such admirable means and appliances for securing our happiness and comfort. Oh! if men would only reflect upon the wondrous power of the Creator, as shown in their very selves, and would *inwardly* confess His majesty and beneficence, is it not probable that a more gratifying offering would thus be presented than in many of the empty conventional mockeries by which we deceive ourselves in an endeavour to defraud Heaven.

> "Lo! the poor Indian, whose untutor'd mind
> Sees God in clouds, and hears Him in the wind."

CHAPTER XI.

The Kidneys, their general and microscopic Anatomy—What they do, and how they do it—Composition of their Secretion—The Solid and Liquid constituents formed in separate Localities—Pyropathy and the Turkish Bath—The latter calculated to increase the per-centage of Heart and Kidney Diseases—Action of the Hot-air Bath on the Circulation is by *no means* analagous to that of Muscular Exertion—Decrease in the Liquid Elements of the Blood by increased action of the Sweat-glands tends to bring about "Stone"—Arguments in support of the Turkish-bath—The Spleen—Thyroid—Thymus—Function of the Spleen.

DURING our wanderings in the human world through which we have travelled so far together, we did not come across the regions known as the kidneys. We have now, however, arrived at them, and mean to inquire not only into their geography, but into their condition as a manufacturing colony also.

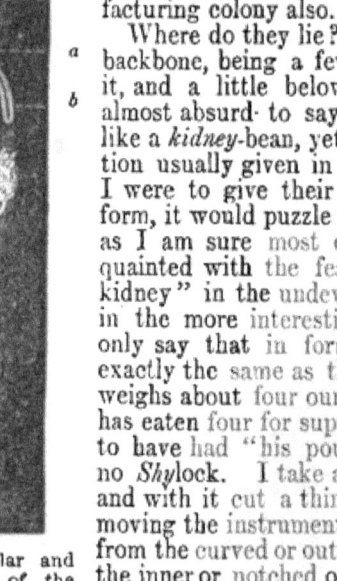

Where do they lie? On each side of the backbone, being a few inches distant from it, and a little below the ribs. It seems almost absurd to say that they are shaped like a *kidney*-bean, yet such is the description usually given in anatomical works. If I were to give their minute characters of form, it would puzzle more than please; and as I am sure most of my readers are acquainted with the features of a "mutton kidney" in the undevilled state, as well as in the more interesting condition, I need only say that in form the human one is exactly the same as that of the sheep. It weighs about four ounces. So that he who has eaten four for supper may be fairly said to have had "his pound of flesh," though no *Shy*lock. I take a double-bladed knife, and with it cut a thin section of the gland, moving the instrument in one direction only, from the curved or outer part of the kidney to the inner or notched one (fig. 51). Now, we have laid out the slice on a plate of glass, and observe that the internal portion is an empty space (*a*), ending in a membranous canal (*b*), whilst the outer or convex part is of a fleshy appearance. On looking at it with a magnifying lens, however, we detect a difference between the outside and

Fig. 51.—Granular and tubular portions of the Kidney, in the upper figure diminished. *c*, the granular, and *d* the tubular parts.

MICROSCOPIC ANATOMY OF KIDNEY.

inside of this mass of tissue,—the first appearing granular (*c*), the second as if made up of tubes (*d*); and furthermore we behold several small arteries making their way outwards, and dividing, and forming knots in the granular division. Next, we remove a little piece of this slice, and placing it between two slips of glass, lay it on the stage of our microscope, and scrutinise it more closely than before. We are well repaid for our trouble, for now the entire structure of the organ unfolds itself to the eye (fig. 52).

We see hundreds of beautiful transparent tubes (*d, d*), expanding at their extremities into little grape-like balls (*a, a*), into each of which a minute artery (*b, b*) enters, forming a knotted tuft within, and from which a *smaller* vein departs.

Let us follow this vein (fig. 53). We see it travelling down by the side of one of the tubes, and then dividing and subdividing, and joining with others, till at length an exquisite network of vessels is produced—an entanglement of meshes—closely surrounding the tube. Glancing still lower (or more internally) the tube is seen, after unit-

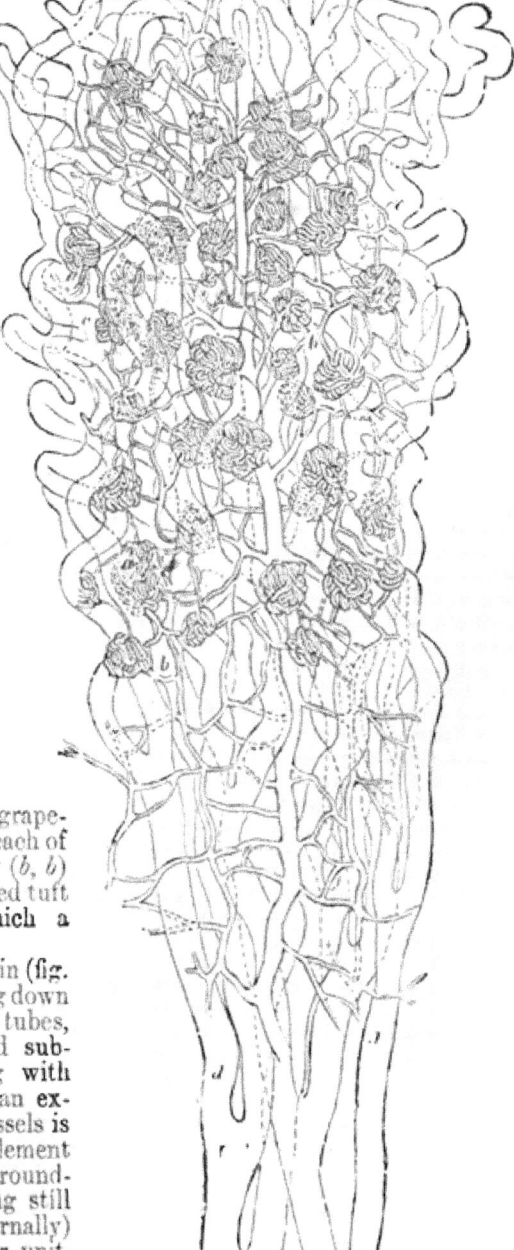

Fig. 52.

ing with others, to end by an open mouth in the empty space which the kidney contains.

The membranous canal that we saw emerging from the notched border is continued onward from each kidney, behind the intestines, to a sac or resevoir which is situate in the lower part of the abdomen, and which in common parlance is styled the bladder (fig. 54).

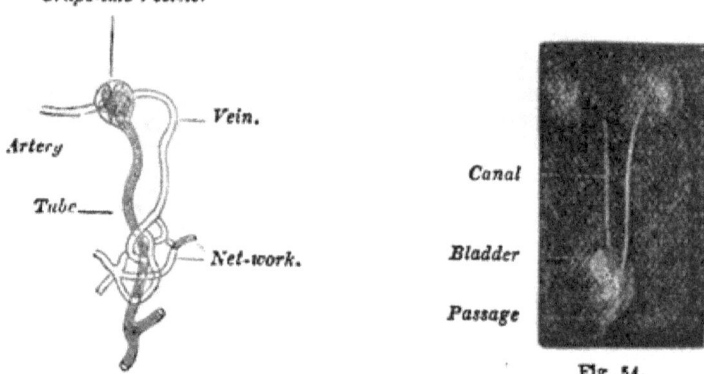

Fig. 53.

Fig. 54.

The duty of the kidneys is to separate a liquid (the urine) from the blood, and this liquid when formed is carried by the two canals to which I have referred, to the bladder; thence, after a period, to be expelled from the body.

What is the signification of all the strange arrangements we have been examining with the aid of the microscope? It is necessary before replying to this question to let you know something of the manufactured compound, before reviewing the processes by which it is prepared.

This liquid is composed of water, holding various salts in solution, which when imperfectly dissolved give rise to the painful disease, *gravel*—or may eventually develop calculus or stone. That beneath is the stereotyped composition:—

Water 938
A peculiar organic substance, urea 30
Salts of potash, soda, ammonia, } 32
lime, and magnesia .. .. }
in 1000 parts.

1,000

Bear well in mind the position of the different structures, and you can attend to the account of the formation of the kidney secretion. The urea and the salts transude through the veins into the lower portion of the tubes, in fact they ooze through; but, on their entrance, they are in too concentrated a form to travel any further, otherwise, as you were doubtless about to say, they would pass into the empty space, and finally into the bladder. How then is their future passage provided for? The watery element of the blood is allowed to flow from the little tuft of arteries contained in the grape-like expansion of the tube, and, as *it* has to reach the empty space also, it naturally pushes before it the crowd of idle bystanders which blocked up the thoroughfare, just as the stream of water from a stop-cock drives before it the particles of filthy matter in a sink or wash-hand basin. What an admirable contrivance, and how wonderfully the means are adapted to the end! Oh, Nephrophagous* gourmand, you whose pleasure it is masticate the renal organs of the ovine genus, yea, even in their most diabolical condition, how little do you think of the thousands of small bodies you are dooming to destruction with that remorseless jaw-bone of yours!

This is the place to give, as Edward Forbes would have said, "a screed anent" the Turkish bath, and its effects upon the system. I take it for granted that you know what a Turkish bath *is*; for, "breathes there a man" who has not heard of Holloway's pills, Du Barry's Revalenta Arabica, and Mr. Barter's public stewing apparatus.† I am inclined to think that one of the most praiseworthy features in the first two preparations is, that, like the physic of the homœopaths, they do no harm; but I unhesitatingly express a conviction of an opposite nature regarding the last. I believe that the constant employment of the Turkish bath is calculated to increase the already too large per-centage of heart and kidney diseases. My belief is based upon scientific grounds, upon experience of the results produced on the health of others by the Turkish bath, and upon the circumstance that this *abominable* system of cooking human beings is happily dying out.

When on the subject of "circulation of the blood," I told you that one of the forces which promoted the conveyance of the nutritive fluid was that produced by the muscular action of the body pressing on the veins, and that when the blood was sent through the system with greater swiftness than usual, a part of the work was performed by the muscles. When you are in the Turkish bath, your blood is driven at a *fearful* pace through your arteries and veins; but there is no extra power developed by exertion, for you are at *rest*. All the labour falls upon the heart, which then not only beats twice as fast as usual, but with more than twice the force, for it has to propel the blood not only through the arteries but through the veins also. And what is the effect of this? The valves of the heart, which are delicate folds of membrane, yield under the undue pressure, and lay the foundation of future endocarditis, disease of the liver, and bowel complaints, horrible to contemplate. But, my dear friend, this is not all—you are not benefiting the physician alone, but the surgeon also. The liquid secreted by the kidneys contains an amount of water sufficient to dissolve up the salts; but is this the case when you have been

---

* From *nephros*, a kidney, and *phagein*, to eat.
† I have been accustomed to designate this form of medical treatment *Pyropathy*.

luxuriating in hebdomadal hot-air baths? No! decidedly, unequivocally no. If I could only reach your ear, and pronounce that negative in an adequately impressive manner, I should be content. While you are sweating yourself (like the jockey who to diminish his weight buries himself to the neck in a dung-hill) your kidney has been secreting its ordinary quantity of inorganic matter—salts; but as you *will* have all the water in your skin, of course this is not *entirely* in a state of solution. The undissolved residue is carried to the bladder, and forms the nucleus of one of those delights of middle age—*calculus*, or *stone!* Beware, I tell you! Picture to yourself that dreadful operation-table—ferocious surgeons, glorying in their *interesting case*—weapons compared to which an Indian's scalping-knives are trifles, and that awful, awe-inspiring un-upholstered and undignified throne—the lithotomy-stool. Again I say, be warned in time! That you may see I am sufficiently impartial to consider both sides of the question, I give the following series of arguments in support of the hot-air bath:—

- Firstly,—It has been employed for centuries by the Turks, who, as is well known, are the most vigorous and energetic people, mentally and bodily, in the world—inventors, philosophers, artists, etc.
- Secondly,—It was never intended that man should employ *mere water* for the purpose of cleansing himself—it being familiar to comparative anatomists that the hot-air bath has been in use among the lower animals since their creation.
- Thirdly,—It accelerates the velocity of the blood, and stimulates the heart to increased action, thus *strengthening* this organ, in the same way as excessive eating promotes a healthy condition of the stomach.
- Fourthly,—It gets rid of all the liquid materials by the action of the skin, keeping the kidneys in reserve for old age, just as by constantly employing only one leg for the purpose of locomotion, when it is diseased we can make use of the other.
- Fifthly,—Thousands of well-educated folk who believe in spirit-rapping, table-turning, and Holloway's ointment, have a profound faith in the Turkish bath also; besides, it is so pleasant to be able to cure one's self and shut the door upon the doctors.

To complete our survey of the glands which are to be found in the body of man, I must say one or two words respecting three organs—the spleen (or melt, of animals), the thyroid, and the thymus. The first is a glandular structure larger than the kidneys, and attached to the left end of the stomach. The second rests against the front of the windpipe, and the last is placed in the chest, behind the breast-bone. It is very difficult to prepare a section of the human spleen; that of the sheep, however, admits of being readily examined under the microscope, and answers our purpose equally well. I regret to say, our knowledge of the function which this organ performs is extremely vague and unsatisfactory, and therefore I shall give in a

SPLEEN—THYROID—THYMUS. 105

categorical form the opinions offered as to its office, without entering into further detail concerning it :—

Firstly,—When the blood cannot flow through the liver, it passes backwards and dilates the spleen, which thus acts as a safety valve.

Secondly,—The blood-disks are renewed here, and waste matter re-manufactured.

Thirdly,—It forms white-of-egg, from the food which has been taken up by the blood-vessels of the stomach, or alters the blood as it passes through.

Our acquaintance with the use of the thyroid and thymus glands is quite as limited; in point of fact, we are utterly ignorant of the offices of all three structures; and when learned lecturers are *demonstrating* in the most lucid manner the functions of these mechanisms, they are not only retarding the progress of science, but deceiving themselves *and* their pupils.

Below are representations of the thymus (fig. 55), and thyroid organs (fig. 56); the latter as seen when a delicate section has been placed under the microscope.

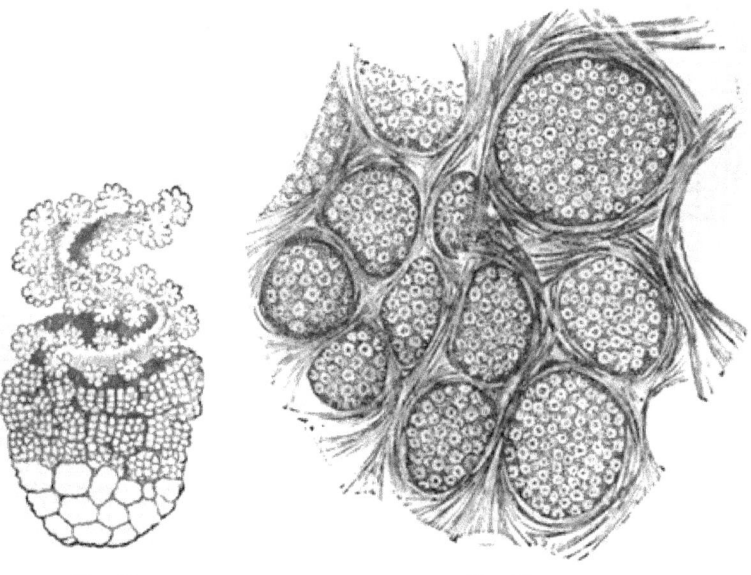

Fig. 55.   Fig. 56.

CHAPTER XII.

Motion and Locomotion—Most Animals can move from place to place, and some Plants have a similar power—The Limbs—The three forms of Levers—Legs and Arms are Levers working on pivots called Joints—Action of a Muscle upon a Limb—Why bodies remain at rest—The Centre of Gravity—We walk by altering the position of the Centre of Gravity and making an effort to fall—Structure of Muscle—Striped and unstriped varieties of Muscular Fibre—How a Muscle contracts.

LOCOMOTION is a faculty possessed in its highest form by animals; although some are denied the power, and certain members of the vegetable kingdom enjoy it. Many of the lower animals are fixed during life to one spot—their existence being a stationary one; whilst some of the most inferior plants are able to move about from place to place with great facility. In one of the earlier chapters I attempted to explain to you the construction of the human skeleton. Here I have to deal with the limbs attached to it, in order to show you the way in which these are worked (in conveying the body from one locality to another, and in carrying food to the mouth), by the agency of numerous bands and cords of flesh, called *muscles*. Do you know that there are three different kinds of levers? First, that in which the pivot, or fulcrum, is placed between the power applied to the lever, and the resistance or weight to be overcome. Of this a familiar example is the common balance, when employed as delineated beneath (fig. 57)—b being the end pressed on, c the pivot, and w the

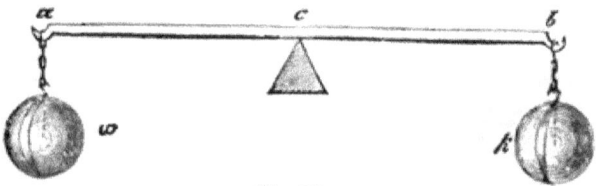

Fig. 57.

weight. Second, where the pivot (c) is at one end, the power ($a\ k$) at the other, and the weight (b) between the two—as at fig. 58. And third, where the weight is at one end (w), the power in the middle (P), and the pivot at the other (F, fig. 59).

The limbs, moving as they do on pivots (joints), constitute a series of levers, whose actions are often complex enough. Let us look at the arm (fig. 60). This is a lever, whose fulcrum is at the shoulder-joint, and which has another lever—the fore-arm (b, c, d) working by a hinge at its extremity. Two sets of muscles clothe these bones—one the flexors, which bend the fore-arm on the arm, and this

FORMS OF LEVERS. 107

on the chest; another, the extensors, which stretch out both, as

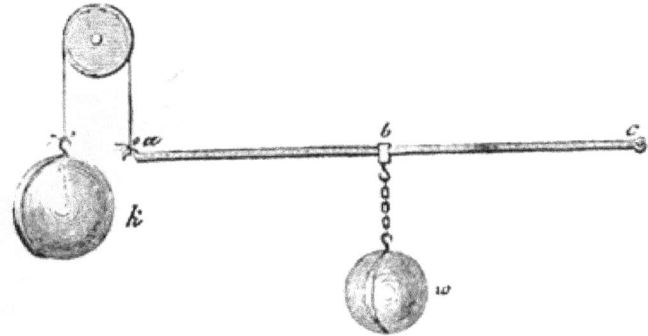

Fig. 58.

when we point to some distant object. The flexor muscle (I only select one) of the arm is attached above to a fixed point—the shoulder; and below is continuous with the bone, through the medium of a strong, unyielding, whitish cord, the *sinew*.

This muscle, when stimulated by the nerves, shortens, and in this way becomes a power (p), operating upon the lever (l), which then is drawn inwards, carrying the weight (w) with it. In the same manner is the forearm (w) bent upon the arm, as shown in the diagram (fig. 61); both levers belonging to the third class. By an exactly similar

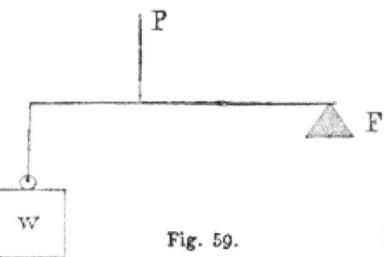

Fig. 59.

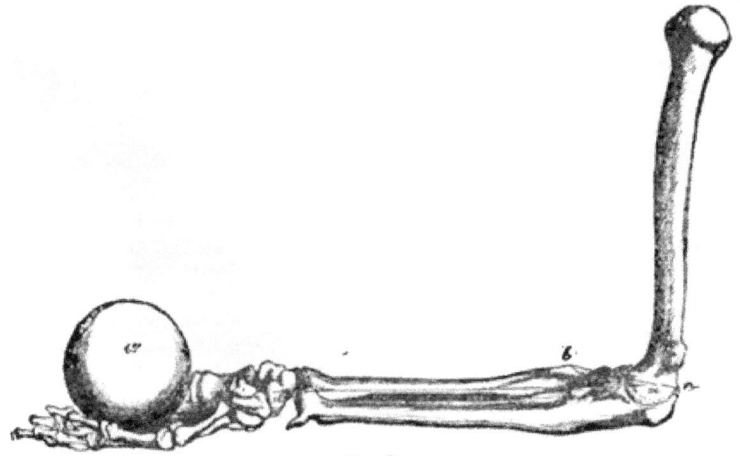

Fig. 60.

set of contrivances on the outer portion of the bones, is the arm extended from the body, and straightened. The head in its nodding movements constitutes a lever of the first kind, or is like the beam of a pair of scales. It rests on two pivots (which practically we may regard as one), and muscles being attached to the skull behind and in front, according as we cause the former or the latter to contract, we elevate or depress the head respectively.

In the thighs and legs the same processes are gone through in moving these limbs. The mode in which the whole body is carried forwards, as in walking, is somewhat complicated. Why does a chair remain at rest when placed on its four legs, and if inclined to a certain extent why does it tumble over? All bodies are attracted by the earth; but in each body there is one particular point, which is attracted, as it were, more than all the rest, in which the attraction culminates; and this is called the *centre of gravity*. As long as any solid body intervenes between this centre and the earth, so long will the chair, or whatever it may be, remain at rest; but the instant it is unsustained, the body naturally falls. Thus (fig. 62):—Here (*b*) the centre of gravity, as indicated by the letter *c*, is placed above the support intervening between it and the earth. If, however, it be inclined till the centre lies outside the support, as in the other instance (*a*), then the body falls. When a man walks, he is constantly throwing his centre of gravity beyond the base of support, and as constantly giving a new support to the falling body. If you have not forgotten the figure of the leg and foot, you'll understand this at once. The leg is attached to the foot by a sort of hinge-joint, which permits it to move upon the latter backwards and forwards. The entire weight of the body is borne by the two feet; and since there are powerful muscles (the calf) uniting the heel and leg, it is clear that when they contract— the toes being the fulcrum, or pivot, the heel will be lifted from the ground, carrying upward, and a little forward, the body. The foot, then, is a lever of the second kind—the fulcrum being at one end, the elevating power at the other, and the weight between both.

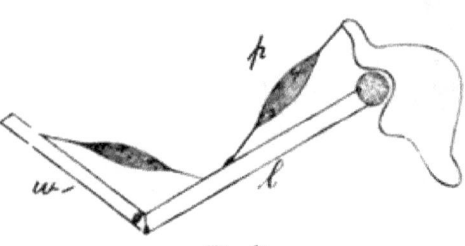

Fig. 61.

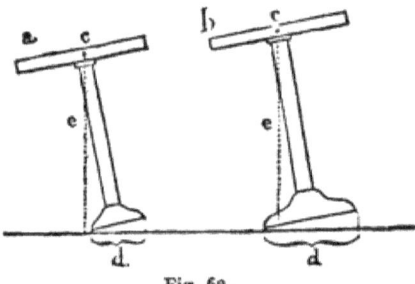

Fig. 62.

How do we walk? When we stand, the plumb-line from the centre of gravity falls between the feet; if we wish to advance, we raise one limb from the ground, and then, by acting on the heel of the other, we throw the body forwards, elevating it a little at the same time. The moment this is done, the plumb-line from the centre of gravity will no longer fall upon the foot, but will lie over the ground in front of this latter, so that the body would instantly tumble, were it not that the second limb is brought down, and thus the centre of gravity being supported, so to speak, the natural state of things is resumed. By a repetition of these processes the individual continues to march onward. Besides the upward and forward movement, there is a slight one from side to side, which is caused by the body leaning a little toward that side, the leg of which is raised; for, the centre of gravity lying in a plane between the two feet, when one is raised the natural tendency is to bring the body downwards. It can't fall toward the supported side, and therefore it does so, in *some measure*, toward the unsustained one. All these different motions combined give rise to the oscillation which we frequently observe exhibited in walking.

Fig. 63.—A man and pack, showing the direction of the line from centre of gravity. *s*, centre of gravity of the pack; *o*, of the body; *g*, common centre of gravity of both; G' g' s', lines from the three centres.

If we were about to investigate into the nature of the various movements of the frame, we should begin a volume especially—so numerous and complex are their characters. We shall therefore now pass on to the consideration of the structure of muscle, and the mode in which it shortens or contracts. Take the biceps—a muscle which every one has heard of—and you will find that it is composed of several smaller bundles of flesh, which in their turn can be decomposed into still minuter ones, and these are made up of what are called *muscular fibres*. Each fibre,* if examined under the microscope, is seen to consist of an outer transparent structure resembling a tube, and an inner series of disks of polygonal form, piled end to end, and of which there are several rows arranged side by side (fig. 64). The disks are separated from each other by

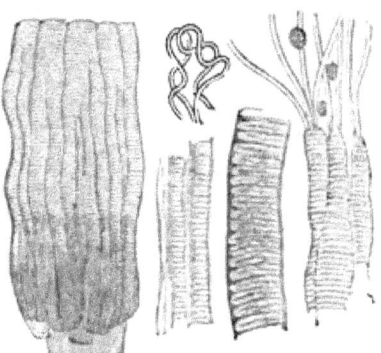

Fig. 64.—Striped Muscular Fibres.

* It is not to be supposed that I bind myself to a belief in the existence of actual fibres; I merely use the term as being convenient for the purpose of explanation.

a soft gelatinous material, and hence, when the so-called fibre is placed on its side, the light coming more strongly through one part than another produces an appearance such as would be observed if the object were actually striated, on which account this form of muscle is termed "*the striped variety*" (fig. 64), although the striping is only *apparent* and deceptive. The muscle contracts by a shortening of its constituent fibres, which diminish their length in the following manner: Each disk widens, and approaches its fellows, at the same time having undergone a diminution as regards the length of the muscle. By this means, whilst each fibre shortens, it also thickens; and this is why a person who is vain of his biceps causes it to contract before he exhibits it, knowing that when in this condition it is more prominent, from its increased width, than when in a state of rest. The subjoined diagrams (fig. 66 A and B), delineating a row of disks during contraction and relaxation, will assist you to comprehend the whole process:—

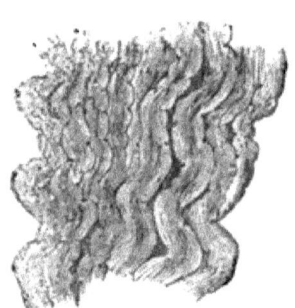

Fig. 65.—Zigzag condition of muscular fibres.

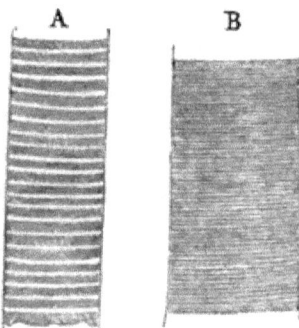

Fig. 66.—Muscular Fibres. A, a fibre relaxed; B, a fibre contracted.

CHAPTER XIII.

The Nervous System—Form and Structure of the Brain—Cerebrum and Cerebellum — Convolutions — Medulla Oblongata — Origin of the Nerves — Use of the Brain — Experiments on Birds proving the Function of the Cerebrum—White and Grey Nervous Matter—Brainmarks of Intelligence — Size— Width — Number of Convolutions — Investigations of Wagner—The Cerebrum covers the Cerebellum, not only in Man, but in certain Apes—Weight of the Human Brain—Function of the Cerebellum—Use of the Spinal Chord—Description of a Nerve—How Nerves end—Eighteen Objections to the Doctrines of Phrenologists — Comparison of the Skulls of various Nations — Long Heads and Short Heads—Straight Faces and Sloping Faces—The Ethnologic Poles—Age of Man—Geological Evidence to prove the Immense Period that has elapsed since Man's First Appearance.

COMPLEX and difficult to be understood as the various organs we have hitherto been considering may have seemed, those which we are about to begin the study of are the greatest puzzle of all; and, indeed, were it not that one could hardly omit their consideration, I should gladly have left them out, for I conceive, that even though I bring forward the most advanced views upon the subject, I shall still be but propagating error; the physiology of the Nervous System, and the nature of life itself being so firmly united, that it is (if possible) extremely difficult to thoroughly understand the former without a clear conception of the latter. And since we do not possess this, it appears to me to be wiser to store up our *facts* till such time as we can explain them thoroughly, than put them before the public dressed and varnished, according to the fashion of pseudo-science.

You will kindly accept the above remarks as an apology for the general

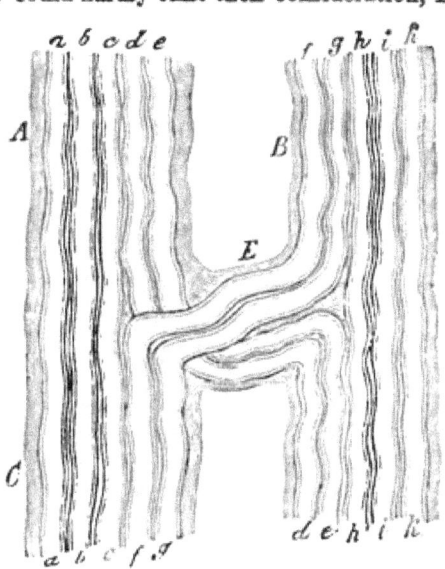

Fig. 67.—Ultimate Nerve Filaments, as seen under the microscope. The fibres of the two sets are seen to cross and intermingle, without however losing their distinctness. The capitals indicate the sheaths of the nerves, the italics the filaments.

112 POPULAR PHYSIOLOGY.

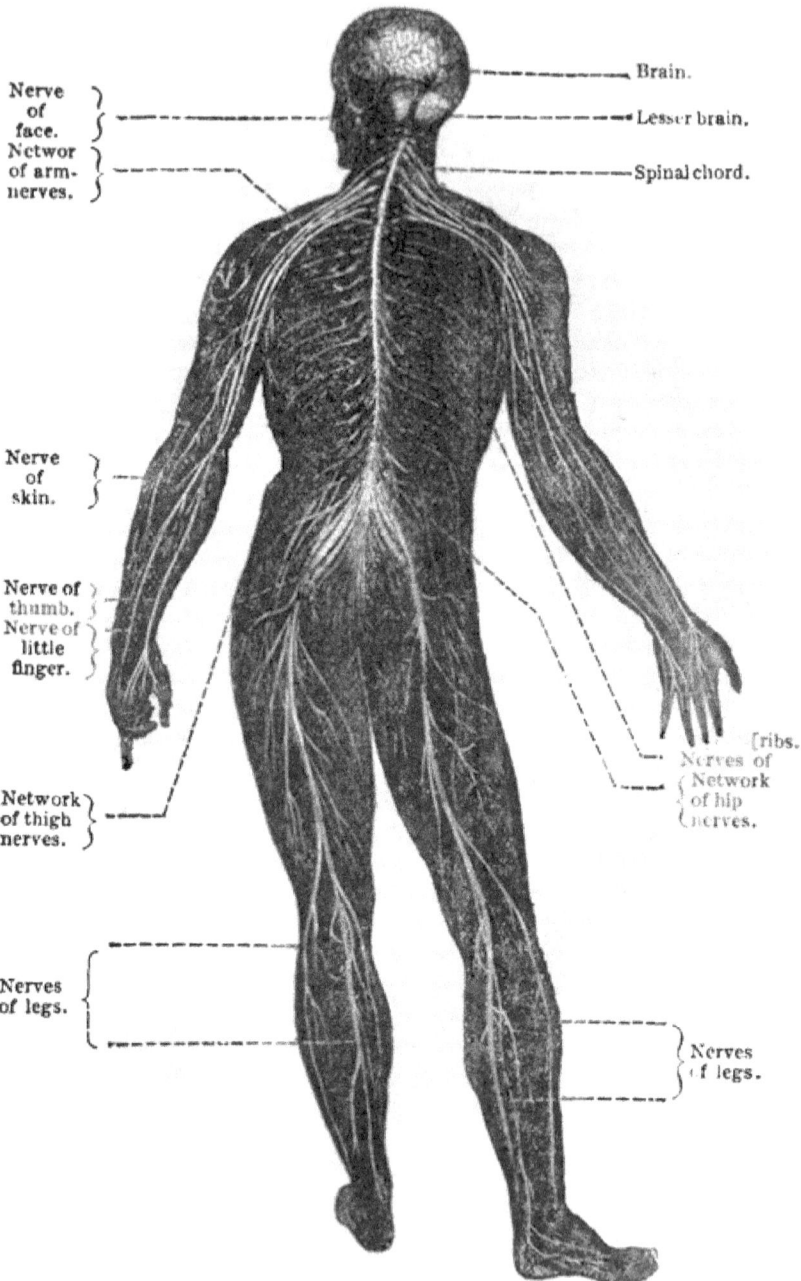

Fig. 69.

STRUCTURE OF NERVE-TISSUE. 113

mode of treatment of the nervous system which I shall adopt in the following pages.

A glance at the diagram will convey to you a notion of the extent and character of the entire nervous apparatus.

It consists of a central chord of a whitish-grey colour, enclosed in the channel formed by the arches of the vertebræ (back-bone), expanding above into the brain, and giving out from its sides numerous branches—nerves—which pass to the limbs and various regions of the body. If we were to place a particle of nervous matter beneath the microscope we should perceive that it was composed of two very different structures—a number of more or less egg-shaped vesicles, and a series of fibres or filaments (fig. 73, A and B). It is thought that the nervous power, whatever it may be (that which stimulates the muscles to contraction,—the glands to secretion,—the mind to the formation of ideas), resides in, and is developed by the vesicles ;* whilst the office of the fibres is to convey this power *from* the vesicles to the muscle or gland, or to convey an impression from the surface of the body to the vesicular substance. The brain and great chord, or *spinal* chord, as it is more frequently termed, are composed superficially of this vesicular material, and hence it is supposed that in these two organs lie the peculiar powers of nervous tissue. If we were to examine the brain minutely, it would form a pursuit for, I might almost say, a life-time, and in the end we should find much of our study flat and unprofitable ; for of what interest could the arrangement of its component parts be to one who was entirely ignorant of their purposes?

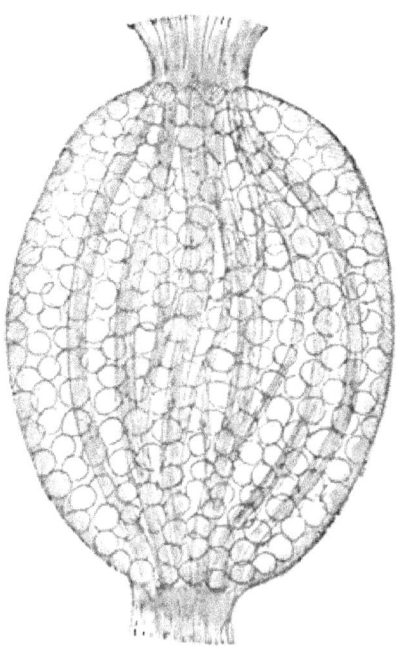

Fig. 69.—Cluster (technically, a ganglion) of nerve vesicles, with nerve-filaments passing through it.

When we look down upon the human brain (fig. 70) we see a whitish-grey, waxy-looking mass, of an oval outline, whose surface is curiously grooved and elevated, presenting many eminences and depressions, which at first sight appear to constitute a kind of labyrinth—these are the *convolutions*, so often alluded to by those itinerant charlatans,

---

* This view is at least questionable.

I

phrenological lecturers. This upper structure is the true brain, or *cerebrum*; it is divided by a central groove, passing from front to back, into two portions or hemispheres, and each of these by cross divisions into three smaller ones, called lobes—a front, middle, and back.* On raising the cerebrum behind we observe, what till then were hidden two small hemispheres, united in the centre, and which are collectively styled the *cerebellum*, or lesser brain. This organ presents no true convolutions; it is attached on the one hand to the cerebrum, and on the other to the spinal chord, the upper part of which is contained within the skull. The cerebrum, or true brain, constitutes about six-sevenths of the entire amount of nerve-substance contained in the cranium (skull), the remaining seventh being represented by the cerebellum.

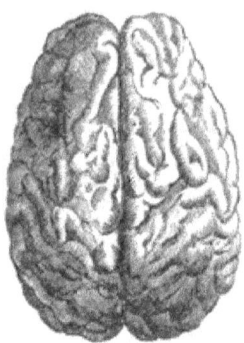

Fig. 70.

The latter, in man, is completely concealed by the cerebrum, and when in its natural position is separated from it by a peculiar tent-like membrane, which is stretched across like an arch from one side of the skull to the other, in the back part of the cavity; upon this the posterior lobes of the cerebrum rest, and beneath it lies the cerebellum.

To that part of the chord, or marrow, which lies within the brain-case, is given the name of "medulla oblongata," and this, as I mentioned before, is continuous with the brain, through the cerebellum. The entire cerebro-spinal system (brain and spinal marrow) is enclosed in three membranes, which lie one within the other; the outer being the strongest and least vascular, the inner the most delicate and abundantly supplied with blood-vessels. This latter covering sinks into the various fissures and divisions of the brain, and in fact has much to do with the nutrition of that organ. The brain is not solid throughout, as we may see by slicing away its substance horizontally, when we come upon two very large, irregularly shaped cavities, of the same form, termed the lateral ventricles; and by making other sections we come across other cavities also. Neither is it a shapeless mass internally, for within the cavities we detect certain prominences and depressions, to which very pedantic names are given, but whose functions, I may safely say, we *know* nothing of.

If we turn to the base, or under portion of the brain, we observe that, like the upper, it has upon its surface many convolutions, which fit into corresponding pits or hollows in the bone. Besides, we see, towards its front extremity, two soft, thin, somewhat club-shaped appendages; these are the "olfactory lobes," or nervous masses which preside over the sense of smell. Resting upon the bones above the

---

* Neither the median groove nor the cross ones extend deeply enough to separate the lobes from each other completely; there still remains a quantity of nervous matter, by which they are united beneath.

nostrils, they send down through the several minute apertures, small filaments to the delicate membrane of the nose. Looking more towards the middle, we remark, placed between the two median lobes, the optic nerves passing from without inwards, and uniting with each other prior to their final separation on their journey to the eyes. Next glancing backwards, we perceive an arched prominence; this is styled the *pons* or bridge of Varolius,* and is composed of a series of fibres travelling from one side of the cerebellum to the other, and beneath which flows a stream of nervous filaments *from* the cerebrum,

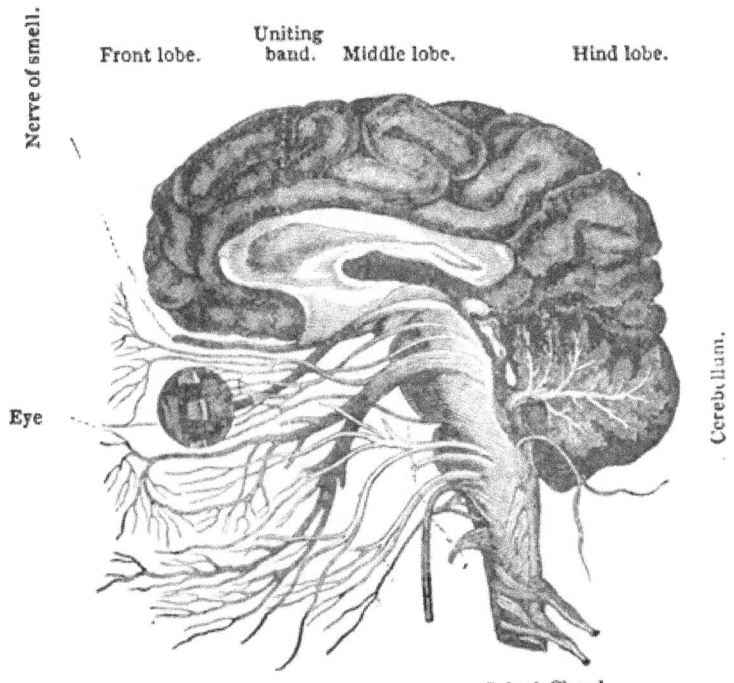

Fig. 71.

which enter the medulla oblongata. This organ is but the extremity of the spinal chord; its most important character is, that within it the rows of nervous cells connected with the nerves of the muscles *cross each other*, those of the right side of the chord passing to the left side of the brain, and those of the left side of the chord to the right side of the brain.†

The spinal marrow itself is divided by longitudinal fissures (one behind, the other in front) into two lateral columns, which together

* A distinguished anatomist of Bologna, died 1578.
† This is the reason why, when the left hemisphere of the cerebrum is injured, the right side of the body is paralyzed.

form a more or less flattened cylinder. A transverse section shows that, though it is white externally, it contains a considerable quantity of grey matter in its interior; this is arranged in the form of two crescents, one in each lateral half; they have their convex borders within, and hence, facing each other—a transverse band of grey substance forming a band of union between the two; their concave edges look outwards, the horns projecting almost to the surface of the chord.

We see, springing from the chord, along its entire length from the skull to the hips, no less than sixty-two nerves, thirty-one on each side; these travel between the vertebræ to the various muscles of the body, and also to the general surface or skin, sending out numerous branches as they journey to their different destinations. If we look at one of these nerves, and observe how it is connected to the chord, we shall observe that it does not become attached as a single string, but that it splits distinctly into two portions, one of which enters the back part, and is *almost* continuous with the hind horn of the grey crescents, the other entering in front, and having *some* connection with the front horn; now these two origins of each nerve are designated the anterior and posterior roots respectively.

There is another portion of the nervous apparatus, the study of which has, till within the last few years, received very little attention. This is the sympathetic system, as it has been called, from the supposition that it exerts a specially sympathising influence over the different organs. The sympathetic system is composed of a number of clusters of nerve-cells (*ganglia*), which are situate in the neck and trunk, on each side of the back-bone. These clusters are connected by filaments, and thus the entire series constitutes a perfect whole. These nervous chains are *directly* connected with the spinal marrow, and have numerous communications with the *ordinary* nerves; they supply branches to the glands, and to several other

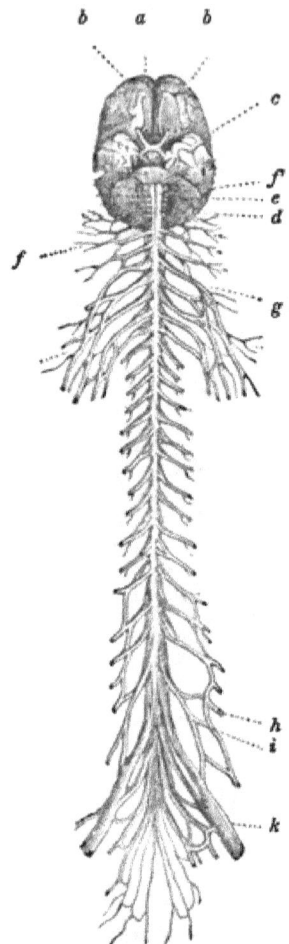

Fig. 72.—Brain and Chord seen from the front. *a*, the cerebrum; *b*, the front lobe; *c*, middle lobe; *d*, hind lobe; *e*, the cerebellum; *f*, the medulla oblongata; *g*, the nerves of the neck; *h*, nerves of hips; *i*, end of spinal chord; *k*, great nerve of thigh.

organs, but their *exact* distribution is not yet understood. It has somewhat recently been shown, that the minute blood-vessels (*capillaries*) are all supplied with filaments from this system; but of this more afterwards.

It seems almost an idle question to ask, What is the exact use of the brain proper? and I should not be at all surprised to hear one exclaim, "Everybody knows that it is the seat of the mind!" Everybody does not *know* that it is the organ of thought; everyone has heard so, and, like many of our so-called *beliefs*, it is the offspring of education. However, both experiment and disease show us that it is the seat of the mind,—of the intelligence, of feeling, thought, and will. When the cerebrum of a pigeon has been sliced away piece by piece,* the animal loses all consciousness, falls into a state of the most intense slumber, and is with difficulty roused by the application of external stimuli; nevertheless, the ordinary processes of the body,— respiration, circulation, secretion, and digestion,— go on as usual; but thought and volition are lost. Thus, if food be placed it its mouth, it will be swallowed and digested, but the bird will never go in search of it; if a pistol be fired close to the head, the sensation produced by the sound will be developed, but the organ of thought being absent, the report of the weapon will not be associated with any idea of alarm, and the animal will remain in the same situation as before. If the membranes covering the *cerebrum* be injured by disease, as in inflammation, delirium supervenes. If the brain itself be diseased, some of the mental faculties are impaired. If the skull, and, consequently,

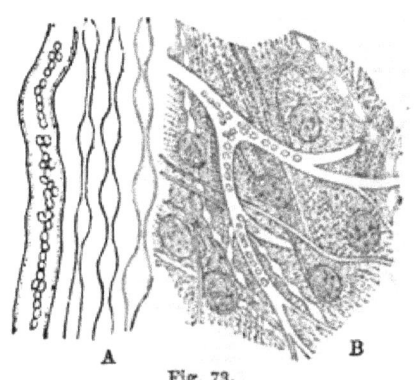

Fig. 73.

the cerebrum, be injured by fracture, serious mental derangement follows. All these facts point to one and the same conclusion, *viz., that in the cerebrum are located the various faculties of the mind.*

The cerebrum is composed of two distinct structures, an internal mass of fibres (?) which are white, and an external stratum of cells which are grey. It is supposed that the latter are connected with the intellectual powers, and that the former are merely the conductors of impressions from without, and of voluntary impulses from within.† I do not think that this has been proved to be the function of the central mass of the cerebrum; but, be that as it may, the idea that

---

* The bird suffers no pain when the cerebrum is being removed, for this organ is quite insensible to torture. The brain of man has occasionally been *incised* without the production of the slightest painful sensation.

† From the body to the brain, and *vice versa*.

the grey matter is connected (in its entirety) with the mental operations, is supported by the following facts:—

First. In infancy, and during fœtal life, when the intelligence is either at its minimum, or nil, the quantity of grey substance is very small.

Second. In idiots there is a proportionally smaller quantity than in intelligent men.

Third. As we pass in the scale of beings from fishes to man, we find the quantity of grey matter increasing.

If the cerebrum of man were simply a convex mass, with a smooth surface, it is obvious, that as the grey matter formed only an external coat, the entire amount could not be very great; but it being necessary for the offices of the mind that a larger surface should be present, we find that in the human cerebrum there is a peculiar arrangement by which the superficial area is vastly increased. This consists in the existence of numerous deep grooves, which traverse the surface in every direction, and whose sides and bottoms are lined with grey matter. If you cannot understand how, by this means, the extent of surface is magnified, fancy that the grey material is a napkin, with which it is required to cover the brain. If, now, it be only required to lay the cloth over the surface, a very small towel will suffice; if, however, it be necessary to allow the cloth to dip into an immense number of grooves, a very large towel must be employed.

We find that, to *a certain extent*, the amount of intelligence is in direct proportion to the quantity of the grey substance, and that, therefore, the animals whose brains are grooved (convoluted) are the more intelligent. This holds good only within certain limits; for example, an animal may have numerous convolutions, and yet, from the fact that the grooves are of no great depth, the amount of surface is smaller than we might have at first supposed, and hence the exhibition of intelligence is also minus.

It is no easy matter to erect any *cerebral* standard of intelligence, when we come to contrast the brain of the apes with that of the human subject. Many writers, however, assert that the following features are indicative of the possession of great intellectual powers: large absolute size; width; number of convolutions; the fact that the cerebrum completely covers in the cerebellum, so that when the brain is looked down on from above, no portion of the cerebellum can be detected; great weight.

We shall consider each of these marks (?) of intelligence *seriatim*, and we shall see that they are at the best fallacious.

ABSOLUTE SIZE:—When the brain is exceedingly small, we not unfrequently find that the former proprietor was an idiot; but conversely, we do not always find idiots with brains below the average size, and very often it happens that, of two fools, he with the bigger head is the greater. Phrenologists talk a good deal of the great size of the brain of Cuvier; but they make no mention of the small size of Sir Walter Scott's, nor of that of Locke's, whose mental powers were possibly quite as extensive as those of the great anatomist.

Moreover, man does not possess the largest brain absolutely; that of the elephant, and also of the whale, being individually greater and more bulky. Neither is it relatively the largest, that of the common sparrow being more voluminous in proportion to the size of the body. Again, if size were to be taken as a standard of intelligence, the degraded Negro should be regarded as possessed of as much intellectual power as the European; for the researches of the most distinguished Dutch anatomists have shown that the negro brain is *quite* as *large* as that of the Caucasian and other races.

WIDTH: It is a popular conception that a man with a large forehead must needs be a man of great mental power; but this does not follow as a logical consequence. Indeed, we find the widest skulls among the inhabitants of the north-east of Asia; but I am not aware that the people of those regions are characterised by a *more than usual* possession of intelligence.

NUMBER OF CONVOLUTIONS: This affords no true key to the mental character of the individual. Many of the lower animals have (according to good authorities) quite as many convolutions as man, and yet hold a very inferior position in the intellectual scale; and certain birds, which have no convolutions at all, are vastly more sagacious than quadrupeds whose brains are ploughed by numerous furrows. It has been stated also, by a most illustrious anatomist, who paid much attention to this subject, that "the brains of some *very intelligent* men were among those poor in convolutions." *

With reference to the position of the cerebellum, as indicative of the presence or absence of intelligence, but little can be said with accuracy. One distinguished anatomist holds, that in *man alone* is the cerebellum completely covered in by the cerebrum, and he therefore looks upon the position of the former in the human brain as indicating the presence of high mental powers; however, many investigations into the position of the *lesser* brain in apes have been made within the last three years, and unfortunately for the superior cerebral character of man's brain, it has been demonstrated that the cerebellum in some of the former animals is also concealed by the cerebrum, and that the difference between the extent of cerebellum covered, in man and in the ape, does not amount to the *one-tenth of an inch*.

THE WEIGHT of the brain of an adult male European varies between 3 lb. 2 oz. and 4 lb. 6 oz.; that of a female varies between 2 lb. 8 oz. and 3 lb. 11 oz. The lesser weight of woman's brain does not show any mental inferiority; for it must be borne in mind that the average female body is much smaller than the average male one, and that, therefore, her brain organs are present in due proportion. If her brain were as large as man's, then by the same line of reasoning she should be the superior being, for her cerebral system relatively to her trunk would be much greater than that of man. Those who assert that absolute weight of brain is a standard of mental force, should recollect that although it is known that the brains of Cuvier, Cromwell, and Dupuytren, far exceeded the average, yet we have little evidence as

---

* Wagner, in the "Göttingen **Nachrichten**" for 1860, No. 7.

to whether very intelligent men may not have light brains, and that ere they give their support to a theory they should have evidence on both sides. Wagner asserts most positively that, *the brains of intelligent persons cannot be proved to be heavier than those of individuals less gifted in mental attributes.* For my own part, I must confess that it seems to me unfair to draw any conclusions from the existence of slight distinctions as to weight, because it is impossible by this means to form an exact estimate of the quantity of *nervous* matter present, inasmuch as no precise deduction can be made for the blood or serous fluid which may be present. Thus, one man may die while his brain has been much excited, and, although he may have been devoid of any great mental powers, his brain, owing to its being charged with blood, will weigh more than it would were the mere nervous matter considered. Again, a man who has been distinguished for his talents, thought, and so forth, may die, and from the fact that previous to his death his mind had been *very* slightly excited, his brain will weigh far less than we should have supposed.

From the above remarks, the reader may conclude that we cannot place *much* reliance upon any *one* of the characters I have mentioned, if we wish to be guided accurately to a fair decision. There are, however, *two* important features in the brain of the European which at once distinguish him from the lower animals. First, *his* brain is larger in proportion to the nerves which emanate from it than that of any other animal; and second (a character which has of late been much dwelt on), his brain is *vertically* vastly higher than that of any existing being. This character of altitude seems to form the great line of demarcation between man and his fellow animals, and is I believe the only one upon which we can place dependence. A great many people of a would-be religious turn of mind have been greatly horrified during the last couple of years by the notion, that "man and the monkey" are in physical features so closely allied. Why they should have been so seized seems to me inexplicable. I conceive that the anatomical characters of man's brain do not explain why he possesses so much intelligence. Let us suppose that mind is entirely the offshoot of brain, then the brain of man has a certain intellectual value ; now the brain of the ape must *similarly* have a constant intellectual value ; and since the distinction between the brains of the two animals (anatomically) is perceived with difficulty, so is the intellectual difference hardly perceptible, a conclusion which (I need not say) is absurd. I think then that the charge of materialism which has been urged against those who elevated the ape *anatomically*, is far more justly applicable to those pseudo-philosophic conservatives who would force on *themselves* the faith in man's *immense* cerebral superiority. From all that we have seen, then, we may pretty safely say that the cerebrum is the seat of the mind, and that in man it differs to *some* extent from that of the ape, but *certainly not in proportion* to the highly developed mental qualities of the former.

What is the function of the cerebellum is the next question to which we shall apply ourselves. As to its office, there are two views current,—an old and a recent one. The first holds that it is the organ of amativeness, and has been especially **espoused by phrenologists**. Scientific men of the present day put no **credence in this doctrine**, for the following reasons:—

First. When it has been removed artificially, or has been **congenitally absent**, the amatory passions have been as intense as **usual**.
Second. In animals which have been emasculated, it is larger **than** in those which are in their natural condition.
Third. It is large in animals whose passions are slight, **and feebly** developed in those whose feelings are very intense.

The second opinion as to its function is more rational. It is thought that the cerebellum presides over muscular actions, enabling a series of movements to be combined in such a manner that a certain end is attained—as, for example, in walking, leaping, dancing, &c. A pigeon whose cerebellum had been removed, though in the possession of all its senses, and of volition and thought, could not move its limbs *in order* (so to speak). It was to all appearance intoxicated—would move one limb first, and quite independently of the other; and similarly with its wings, so that it stumbled about without advancing. This was, of course, an experiment, produced by artificial means; but disease also supplies us with experiments, which are often more valuable than those conducted within our laboratories. When the cerebellum is diseased the same want of co-ordinating power is observed that was exhibited by the pigeon. A man afflicted in this manner can make no *complex* movements, and occasionally is seen to spin violently round and round in one direction, or to roll over and over upon the ground. In what particular manner the cerebellum performs its work we cannot tell; it is one of those mysteries of the nervous system which it will **require** the labour of centuries to unravel; all we *can* say is this: **it holds** the same relation to the different muscles of the body that **the** conductor of an orchestra does to the musicians. In both cases, were there nothing to regulate, as it were, the time, each of the many units of the entire combination would act irrespectively of the others, and instead of a harmonious result we should have but discord and confusion.*

The oblong marrow, or *medulla oblongata*, as it is more technically styled, has two uses: it conveys the cell-rows of the spinal cord to the cerebrum and cerebellum, and is said to preside over the function of respiration; for, when it has been punctured or destroyed, complete asphyxia or suffocation **ensues**. From it spring the important nerves which supply the lungs, larynx, and stomach; and it contains within it a small centre of grey matter, which is named the "*vital spot*," because when *even touched*, life ceases. This fact, when we have given the subject no consideration, appears extraordinary; but when we know that nerves supplying the larynx, lungs, heart,

---

* It must be here stated that **even** this view of the function of the cerebellum is open to the very serious objection, that in some cases of cerebellar disease the muscular movements are unimpaired.

stomach, and liver all take their rise in the vicinity of this particular region, much of our surprise is lost.

The spinal chord has a two-fold office. First, it serves to convey *to* the brain various impressions which have been received at the surface of the body (skin, fingers, etc.), and carried to it by the nerves of sensation; and also to convey *from* the brain the impulse of the will to those nerves, which, passing *from* the chord, are distributed among the muscles. Second, it operates as a sort of animal brain, possessing, as it were, a kind of consciousness, which enables it to superintend the working of such organs as the gullet, without the interference of the true brain. Thus, when we sleep, it is requisite that certain actions be performed—as, for example, the contractions of the diaphragm, the elevation of the ribs, &c., &c.; and if these offices were placed under the direction of the will, it is evident that on falling asleep we should forget them, and that we should therefore be suffocated. Now the spinal chord (being, in a manner, a brain), being aware of the necessity for the performance of these functions, stimulates the organs, and these discharge their duties, even though the chord be completely severed from the true brain. To this process, in its entirety, the term reflex action has been given, to distinguish it from the ordinary operations of the animal, which are the result of volition. As we shall now see, the term is likely to lead us astray. How is this reflex or reflected action produced? The lungs (if respiration be interrupted) become charged with venous blood, and produce a peculiar impression, which is transmitted to the upper part of the chord (or possibly directly), and from it to the middle portion, which then, by influencing the nerves proceeding to the muscles, causes them to contract, and in this way recommence the respiratory process, which had been interrupted. In this case some suppose that the nervous influence transmitted to the chord was *reflected* from it to the muscles; but this is purely a supposition—the process bearing, on the whole, a great resemblance to our ordinary voluntary acts; these being invariably the result of either an impression generated in the mind, or conveyed to the nervous centre from without, and might with some show of reason be called reflex. In neither case is the nerve force *reflected*, but in both the centre is stimulated, by some impression, to develope force, which is then borne along the motor nerve to the muscle.

I mentioned before that the nerves enter the chord by two roots: a front and back. Now, these roots have very distinct properties—the front being connected with motion, and the back with sensation; therefore, as these combine to form a single nerve, it is evident that each nerve is compound—that is to say, it contains some fibres whose office it is to convey impressions to the chord, and others which serve to carry a motor impulse *from* the chord *to* the muscles.

A description of one nerve will suffice for all. It leaves the vertebral column as a single whitish chord, but as it travels towards the surface of the body, it divides and subdivides, sending off numerous branches to the different muscles and such like, and eventually terminates in the skin. The quantity of ultimate nerve fibres contained

DISTRIBUTION AND TERMINATION OF NERVES.

in any nerve is almost immeasureable (a single fibre rarely exceeding one four-thousandth of an inch in diameter); and since each fibre is capable of conveying an impression distinct from any of its fellows, some idea may be formed of the perfection of the whole arrangement.

The sensitive fibres are attached *principally* to the skin, the motor to the muscles. Why? Because the muscles being organs of motion they require nerves which will convey the nervous stimulus *from* the chord *to* them ; and the skin being the part upon which external objects impinge, it requires nervous fibres capable of transmitting the impressions received by contact, *from* it *to* the chord and thence to the brain, in which, by the assistance of the mind, they are appreciated.*

How the nerves terminate is not well known. Some say in loops; others, that they end abruptly; and a recent writer asserts that in the muscles they are continuous with microscopic bodies called corpuscles. It is *known* that in many parts of the skin (fingers, &c.) they end in peculiar, solid, crow-top-like structures, designated Pacinian bodies.

A little reflection will unfold to the reader the different parts which all the organs I have been describing play in the living body; and therefore with the subjoined diagram, explanatory of the mode by which sensory and motory impressions are conveyed, I conclude my remarks upon this branch of the subject.

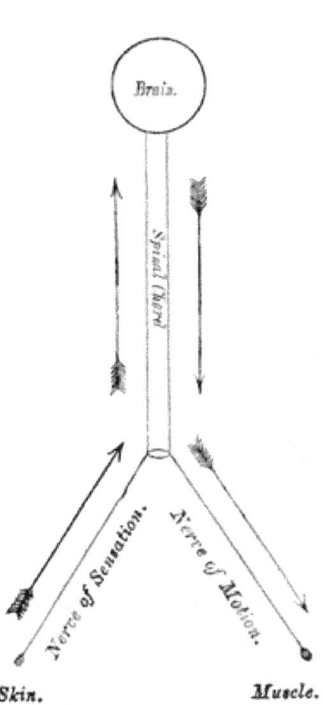

PHRENOLOGY.

As this is a subject to which much attention is given by the uneducated classes of society, and by a few also of those who have had opportunities of knowing better, but who love to be told by some nomadic vendor of characters, that they are likely to become Shakesperes, Miltons, Newtons, and so forth, I think I may

---

* It is to be regretted that, at a meeting of the British Association two or three years since, a physiologist, in communicating some exceedingly original views upon the distribution of motory and sensory nerves, confounded these with the spinal roots of *admittedly compound* nerves, and has thus led his unscientific readers to form many misconceptions.

conscientiously say a few words about it. To the credit of Anglo-Saxon logic, it must be stated that the science of "bumps" did not originate in this country, but in that which of all others is renowned for its fanciful generalizations. There is no need to enter into the doctrines of phrenologists, for to those who are already familiar with them my objections will be sufficiently explicable, and to those who are ignorant of them I would merely say, *that* "ignorance is bliss," for phrenological wisdom is assuredly the extremest folly.

Objections to the Doctrines of Phrenology.

1st. It has never been *proved* that the human mind is composed of the several faculties which phrenologists enumerate.

2nd. It has never been *shown* that any of these faculties reside in one particular convolution, and in no other.

3rd. If intelligence be in the grey matter, as phrenologists admit, the involutions (foldings inward) must have faculties as well as the convolutions; yet as these sink into the brain, no cognizance is taken of them, though the quantity of grey matter situate in them is vastly greater than that in the convolutions.

4th. The most extensive convolutions are seated in the interior and on the base of the brain, and are consequently never recognized by phrenologists; but these should possess faculties also, and should therefore be considered (which is impossible) in forming an estimate of one's character.

5th. The brain may be elevated at some point of the surface, and phrenologists would say the faculty of this particular region was well developed; but as the superficial part might not have been developed, but simply pushed up by the development of some deep-seated structure, an error may often be fallen into.

6th. Faculties are placed *beside* each other, though without the remotest affinity; hence we might reasonably expect that there would be in the brain a distinct line of demarcation between the convolutions in which these faculties lie. But such is not the case; one convolution passes insensibly into the other, like a wave on the sea; so that it is *utterly* impossible to state exactly where one begins or the other ends.

7th. There may be a large brain, with large convolutions, yet from the fact that the *involutions* are shallow, the entire extent of surface will be smaller than that of an average brain, and consequently the intellectual power must also be less.

8th. I have already shown that the cerebellum does *not* preside over the amatory passions, as is supposed by phrenologists.

9th. If it be true that each convolution represents a distinct faculty, then is the common porpoise as highly gifted (in regard to the number of its faculties) as man.

10th. Memory *may* be lost by injury to *any* part of the brain.

11th. When portions of the brain have been removed, the mental faculties have remained unimpaired.

12th. The convolutions are *not* symmetrical; for example, the convolution representing "hope" is not placed exactly opposite its fellow.

13th. The instinctive faculties are possessed in the *highest* degree by the inferior animals (invertebrata, as insects, &c.), which have no brains.

14th. Although internally the skull forms a sort of cast of the convolutions, yet *externally* this is *certainly not the case*. The skull is formed of two layers of bone, with a spongey substance between them; the inner one maps out the elevations of the brain, but these latter make no impression on the outer one; so that where there is an actual depression of the brain there is often an external elevation of the bone, and where the brain is elevated it sometimes happens that the outside of the skull is flattened.

15th. A case has been lately recorded of a man, one of whose hemispheres was completely destroyed, and who, nevertheless, possessed his intelligence intact.

16th. M. Esquirol, who has had, perhaps, the greatest opportunities of testing the truth of the phrenological doctrines (having been for many years connected with the most extensive lunatic asylums in France), has come to the conclusion, that in no case is phrenology to be depended on in seeking an estimate of men's characters.

17th. The remark of Napoleon to Las Casas may be cited as a fair objection, although there is no bump of drunkenness. Speaking of Gall, he observed, "He ascribes to certain prominences propensities and passions which do not exist in nature, but are the growth of society and merely conventional. What would the organ of theft effect if there were no property; the organ of drunkenness, if there were no spirituous liquor; or the organ of ambition, if there were no society?"

18th. If we were to cut down through the portion of the skull beneath which phrenologists suppose the faculties of colour, order, and number to lie, we should meet with a large space between the outer bone and the inner, partly filled with loose spongy tissue; thus proving that the prominence does not indicate any increased development of the parts of the brain which are placed beneath it.

If the foregoing statements, the correctness of which is undeniable, do not at once banish from the reader's mind every vestige of a faith in phrenology, I shall think that conviction is impossible. It is exceedingly remarkable, that the "amatory function" of the cerebellum was a fact (?) in phrenology which Spurzheim believed was supported by a *more overwhelming mass of evidence* than any other. Alas! alas! poor phrenology! If these be thy surest props, what becomes of thy other supports? Must we consign thee, in fellowship with mesmerism, electro-biology, and spirit-rapping, to the hands of that Barnumizing professor of bunkum-science, the wandering Yankee lecturer?

Ere we close this chapter, "**one last remark I wish to make, one** last explanation I wish to offer," in connection with the skulls of men, and the period which must have elapsed since man first made his appearance on the globe. The skulls of the different nations of the earth, though varying immensely, may yet be divided into two classes, — the long-headed and the round-headed. In the first, the length from forehead to poll greatly exceeds the transverse width; in the second, the difference is less marked. In the long head the breadth is usually about six-tenths of the length; whilst in the round head the transverse diameter is, as a rule, about nine-tenths of the "fore-and-aft" measurement. We find the round heads best represented among the Turks and Tartars, and the long heads among the negroes, whom I fear we cannot call long-headed in any other sense. This difference in the conformation of the skull, however, does not point to any ethical difference between the characters of the races possessing these skulls; for, as a distinguished palæontologist observes, "we may find two races, such as the Caffre and Calmuc, with very opposite measurements, but yet very similar in morals and manners, or rather in the total *absence* of both morals and manners." Besides this distinction as to length and breadth, there is another and important distinction, with which it is associated. In some faces, such as those of the negroes, we see the face and jaw-bones projecting in advance of the forehead; in others, on the contrary, we perceive that the face forms a less or more continuous right line with the forehead, as in Europeans. To the first form the term "prognathous" is applied, and to the second the term "orthognathous;" and it has been found that the prognathous face and long skull are companions, and that the orthognathous face accompanies the round head.

From an examination of the skulls of the different peoples of the world it has been found possible to draw a sort of ethnological meridian with its two poles. Thus, if a line be drawn across the map of the world from Russian Tartary to the Bight of Benin (on the south coast of Guinea), and starting from its centre we pass towards its north-eastern extremity, we shall find ourselves among races more round-headed and straight-faced as we approach the pole. If, on the other hand, we steer southward, we shall pass over a country inhabited by races with long heads and projecting faces, until we arrive at the southern pole, where these characters are most prominently represented. This law is possibly better established than any other in the entire science of ethnology, and is pregnant with interest to the philosophic mind.

When I ask, what is the length of time during which man has existed on the globe? I have no doubt that many of my fair readers will reply, about five thousand five hundred years. Now it would ill become me to endeavour, by contradiction merely, to shake one's faith in generally assumed truths; but facts are never polite; there is no "by your leave" or "begging your pardon" among them, and as everyone has a few stubborn friends, who will "speak their minds" in true *Brummagem* fashion, I trust I may be excused for introducing some very obstinate fact-acquaintances of mine.

Before doing so, however, I must premise that archæologists admit that, prior to the great Roman nation, there lived upon the globe two distinct races of men, one of which succeeded the other. The first (second in point of time), relics of which have been found pretty frequently, were a long-headed people, and were workers in iron. The second were workers in bronze, but not in iron, possessed some cultivation, and domesticated animals for their use. This last race is the earliest to which *archæologists* can point.

Beyond these, however, there was another race, the bones of which have been found associated with weapons of stone and flint, but without any remnants of domesticated animals. To these races have been given names taken from the materials in which they worked. Thus the first has been called the Iron, the second the Bronze, and the third the Stone race. Now for our facts: "Denmark possesses great peat bogs in various parts of the country, in which are embedded forests of trees. In the more superficial layers of the soil are embedded fallen trunks of beech-trees—great trunks of beech, like those which now adorn the surface of the country, and are its chief and most graceful decoration. Beneath these beech-trees we come to a lower forest—a forest of oak-trees, fallen, with their tops to the centre, of noble size—oaks which had taken centuries to grow, and have been centuries in the ground. Dig deeper again, and you come to another forest of large and splendid pine-trees—noble trunks of three feet in diameter, of great age and magnificent proportions. Now, in the memory of man, *there has been nothing in Denmark but beech-trees*. Past the memory of man grew and flourished those giant oaks; they had centuries of growth, and for centuries they have been buried. Past these centuries we must look down through the vista of ages for the time when the pine-trees stood erect, and slowly gathered their bulk, and fell into the lowest part of this deep peat, to be again covered with the wrecks of succeeding epochs of vegetation." "Now, in the beech forests of the bog we find only traces of the men of iron; amongst the oaks, only of the men of bronze; and amongst the pines, only of the men who worked in stone. Beneath the pines we find only peat, and no remains of man of any kind whatever."

"What is meant by this lapse, not so much of time as of facts? *We cannot number the ages* that saw the rise and fall of these monarchs of the vegetable world and the succession of these races of men."*

The foregoing paragraphs speak so strongly for themselves that any additional comments of mine would only detract from their force. There is an old German maxim which in this instance and in all similar ones is worthy of adoption, and as my library *does not* contain a copy of Tupper I shall, in terminating this chapter, quote it in the original:—

"Reden ist Silber, aber Schweigen ist Gold."

---

* A lecture on the "Fossil Remains of Man," delivered at the Royal Institution, February 7th, 1862, by Professor Huxley, F.R.S.

CHAPTER XIV.

Organs of Special Sense—The Eye—Laws which regulate the Passage of Rays of Light—Refraction—Effects of Convex Lenses on Light—Spherical and Chromatic Aberration—Form of the Eye—The Coats and Humours—How we see an Object—Why the Image is inverted—Why Bright Light and Darkness confuse us at first—How we focus the Eye—Short-sight and Long-sight—Colour-Blindness—After looking at the Sun we behold a black Disk before the Eye—Why with two Eyes we see but one Object—The Stereoscope—Ideas of Size, Surface, Form, Distance, etc.—Eyelids—Tears—Tear-glands—Diseases of the Eye.

HAVING done with the nervous system, viewed as a whole, and having examined the cerebrum, or organ through which the mind operates in exhibiting the various mental faculties, we come now to the consideration of those mechanisms in which the so-called *senses* are located. These mechanisms are the channels through which *special* impressions travel on their way to the brain. Thus, an impression of light is produced upon the inner portions of the eye, and from these passes to the brain, giving rise to an idea of the object seen. Likewise, in the cases of hearing, taste, smell, and touch, impressions are conveyed to the cerebrum, and ideas in connection with these impressions are developed. In this manner we may regard the brain as a sort of mental stomach, with which are connected a series of mouths (organs of sight, smell, taste, touch, and hearing). Through these mouths the crude food which is taken in is carried to the brain, and is there in the form of ideas, of a loose, heterogeneous type. The brain (through the assistance of the mind) associates these mingled notions, arranges them in bundles, and thus the mental power is increased by (as it were) a species of cerebral digestion. We have now arrived at the senses, and before we commence the study of any one of them it behoves us to inquire whether there is any character common to all, so that we may catch *it*, and form a general idea of a sense. There *is* a common feature, *viz.*, each sense-organ consists of two very distinct portions, which may be termed the *receptive* and perceptive regions respectively. In the ear we find a complex apparatus for the reception of the vibrations of the air which produce sound; and, in addition, we observe an arrangement of a nervous nature, by which the vibrations are *per*ceived, and then carried to the brain. The same thing may be said of the tongue, the nose, the skin of the fingers, and the eye, of which we propose to investigate the structure and function in this chapter.

We say this man sees quite distinctly, but that poor fellow is blind. The one can make his way onward without coming into collision with external objects, the other cannot. Again, no man can walk without danger, in perfect darkness, or with his eyes closed.

These facts prove two or three important things; they show us,

Firstly, that we only become cognizant of the existence of external and surrounding objects through the action of the eye; and—
Secondly, that the eye recognizes these objects through the influence of light.

Therefore, in order fully to understand the mode in which vision is effected, we must make ourselves acquainted with the structures of the eye, and also with the properties of light and the laws by which its rays are governed; in fact, we must learn a little of that difficult branch of knowledge—optics.

Light passes from the body which emits it in straight lines. A ray of light travelling from the sun to the eye pursues a rectilinear course. If, however, on its journey it has to pass obliquely* through some medium of a density different from that of the atmosphere, it will have its path broken, and will travel to the earth in another direction. This process by which the ray is diverted is termed *refraction*.

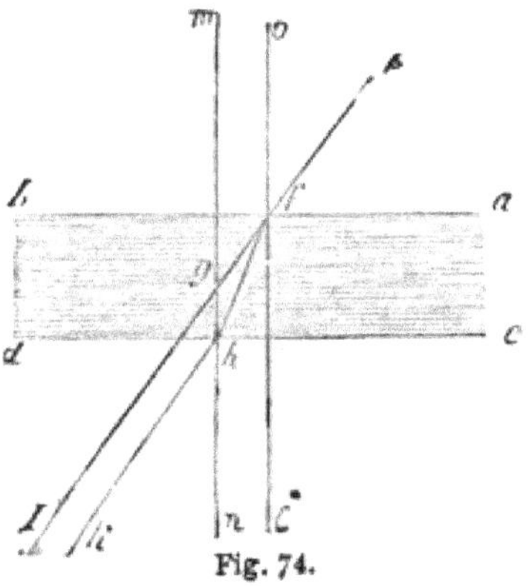

If a ray of light ($p f$) which has travelled from the sun strike obliquely a plate of thick glass ($a b c d$), instead of passing onwards directly in a line ($f g l$) with its path from the sun, it will be turned

Fig. 74.

more to the left ($f h$), and so will travel through the plate till it reaches the other side (h); here, as it emerges from the glass, its course is again changed, and it is bent towards the right ($h h'$). If, then, a perpendicular line ($o i$) had been drawn at the point of entrance of the ray, we should have found that this latter was bent towards it as it travelled through the glass, and from it on emerging.

It is upon this known property of glass (and such-like dense transparent substances) that the formation of images by lenses depends. You know that if a lens (double convex) be fixed in a hole in the shutter of a window and the shutter closed, a sheet of white paper placed at some distance from the lens will have formed on it a distinct picture of everything which is in front of the window. This is explained as follows. Let the arrow $a b$ (fig. 75) be placed in front of the lens $c d$, and let $e f$ represent a screen placed behind for the reception of the image. The rays of light passing from the ends of the arrow will impinge upon the edges of the lens, and will, by refraction, be

---

* Rays which fall vertically upon a medium, such as glass, are transmitted in the same line, and consequently suffer no refraction; but other things being the same, the more obliquely a ray falls (on a lens, for instance), the greater will the refraction be.

K

caused to converge as they leave these edges; therefore they will meet it at a certain point. But this is not all. They then cross each

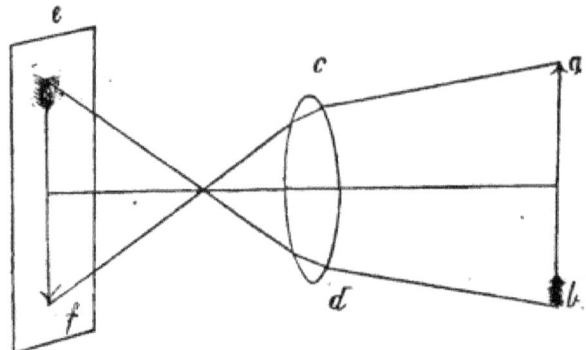

Fig. 75.

other, diverge, and at length form an image (which is *inverted*) on the screen. That the image is inverted will be readily believed by all who understand photography.

The reason why a lens is so shaped, is in order that the rays which proceed from the edges of an object may be caused to converge, and thus all the rays be brought to one point, which is called the *focus*. If it be not at first clearly understood why the rays converge, let lines be drawn perpendicular to both surfaces of the lens, and then let it be remembered that oblique rays, when passing into a denser from a rarer medium are bent towards the perpendicular, and when entering a rarer medium are bent from the perpendicular, and the process will at once be intelligible.

It sometimes occurs that all the rays of light coming from an object do not meet at *one* point after they have travelled through a double convex lens; the rays which fall upon the edges converge at a greater distance from the glass than those which traverse the central portion; hence there results a confused image, and this defect is termed "*spherical aberration*." It is corrected by having the surface of the lens curved in such a manner that all the rays are caused to meet in one focus.

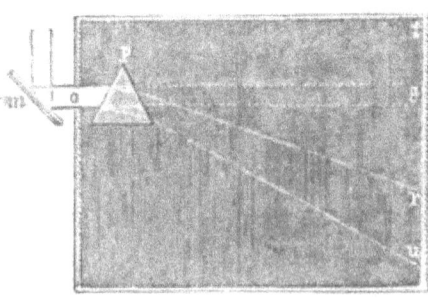

Fig. 76

There is another defect, which many lenses exhibit, and which is called "*chromatic aberration;*" and which consists in the display of various colours besides those of the object from which the rays of light

SPHERICAL AND CHROMATIC ABERRATION. 131

travel. In order to explain this phenomenon I must tell you that white light—that of the sun, for example—is really composed of three distinct colours: red, blue, and yellow. We can prove this in the following manner. Let a stream of light (fig. 76), reflected from a mirror

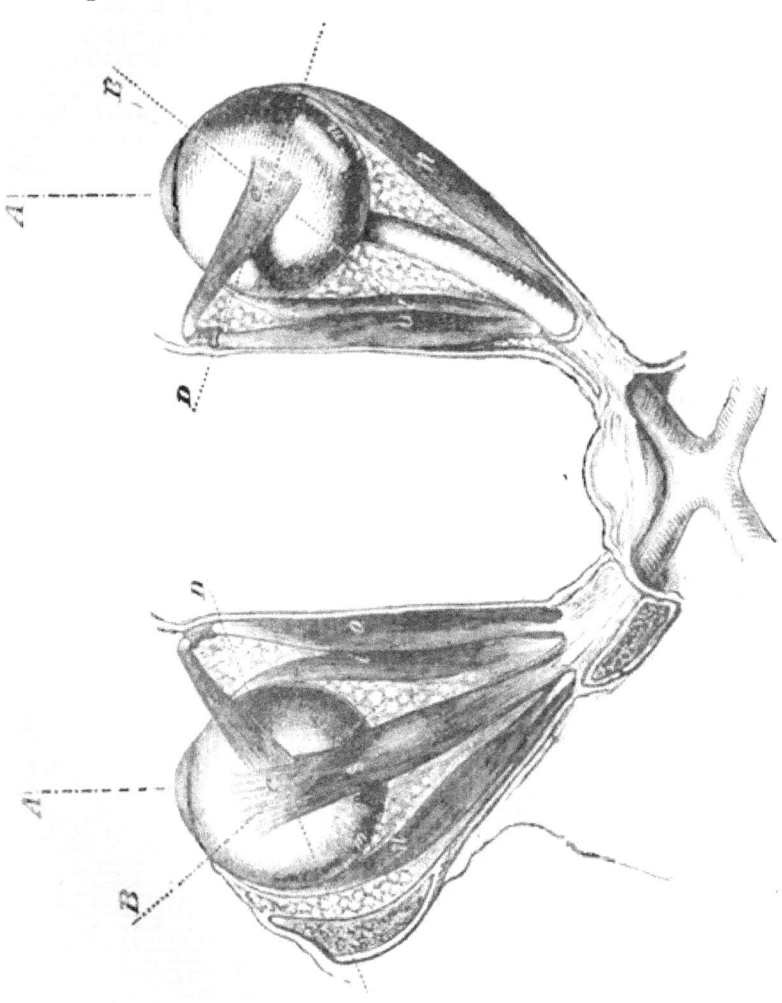

Fig. 77.

(m), pass through an aperture in the side of a dark chamber, and it will produce a circular spot of white light (g). If, however, a glass prism (p) be interposed, the rays will be refracted towards *r u*, and as the colours which together constitute white light are broken

K 2

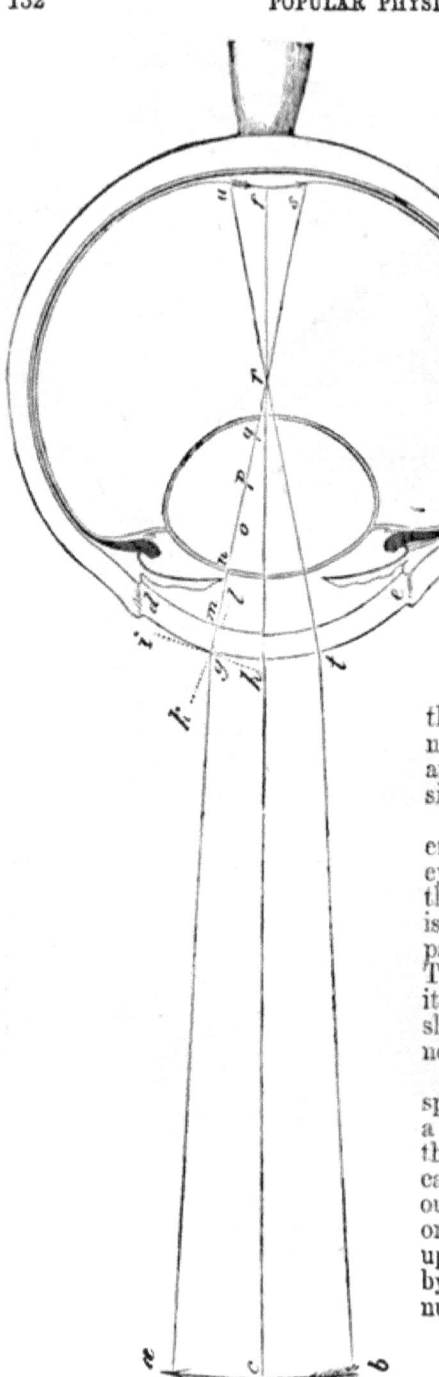

at different angles, and thus separated from each other, we shall have at *r u* a spectrum or row, which, like a rainbow, contains the primary shades; these latter being called prismatic colours, because produced by the action of a *prism*.

Now this property of prisms is also possessed by many lenses, and hence with such, true images cannot be formed. This "*chromatic aberration*," or tendency in glasses to decompose white light, is corrected by combining together lenses of different densities, which are then called "*a-chromatic*," or non-colour-forming, and these are invariably the more expensive forms.

We are now in a position to enter upon the subject of the eye itself. It has been said that the human organ of vision is the most perfect optical apparatus that can be conceived. This is not the case; for although its defects are exceedingly slight and unimportant, yet, nevertheless, they *are* defects.

The eye (A, fig. 77) is of a spheroidal form, and is fixed in a niche or socket, formed by the union of several bones. It can be moved either in or out (*n i*), up or down (*s*), or may be made to rotate upon its front-to-back axis (*o e*) by means of muscles, six in number, which are attached to

COATS OF THE EYE.

its outer surface, and also to the bony socket. It is composed of several coats of membrane, which overlie each other like the skin-laminæ of an orange, and of certain lenses, or humours, which are contained within these coats. The optic nerve enters it behind, and spreads internally in the form of a layer which is termed the retina. I may here mention that the optic nerves, prior to their entrance into the eyes, decussate, or cross each other, and in doing so allow their fibres to intermingle, so that each eye contains fibres from the two nerves, as shown in fig. 77.

The outer surface of the eye is white, save at one place in front, which is beautifully transparent, and through which the rays of light travel on their way to the inner portions of the ball. The wall of the eye is composed of three coats, named from without inwards, the *sclerotic*, the *choroid*, and the *retina*.

The sclerotic, or, as it is more frequently styled, "the white of the eye," is a tough, dense, fibrous tunic, which entirely invests the eye, except at the locality alluded to—the *cornea* (A A, fig. 79). It constitutes about four-fifths, and the latter structure the remaining one-fifth, of the entire surface of the eyeball. Its beautifully smooth and glistening appearance is due to a delicate layer of mucous membrane, which clothes its entire surface, is continued over the cornea, and also over the internal surface of the eye-lids.

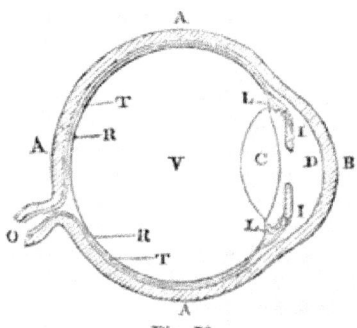

Fig. 79.

The cornea (B)—which we may regard as the clarified transparent portion of the sclerotic—bulges out a little from the eyeball. It is, in fact, a sort of watch-glass, fitted into the sclerotic, which contains a kind of groove into which the cornea is inserted. It is composed of different layers, two of which—the inner and outer—are very elastic. As these are of more interest to the miscroscopic anatomist than to the physiologist, we shall not dwell upon them further. No blood-vessels can be discovered in the cornea; indeed, it is evident that if blood-vessels were present, much, if not all, of its transparency would be lost. But although **no** channels large enough for the conveyance of the red particles **of** the blood exist, there are minute passages through which the purely fluid elements may be conveyed.*

* This has been **questioned by** some writers.

The choroid coat (T T) is composed of hundreds of blood-vessels, and minute polygonal bodies containing black pigment or colouring matter. It lines the whole of the internal surface of the sclerotic, from the entrance of the optic nerve behind, to the junction of the former with the cornea in front, and is continued over the back part of an organ we shall come to presently—the iris (I I). The use of the choroid is by some considered to be to preserve the warmth of the retina, which lies within it. This is not its only object; its black pigment absorbs all the loose rays of light, thus prevents their reflection from one side of the eye to the other, and therefore tends to render images less confused. It fulfils the same office as the black lining to the tube of a microscope, and when absent—as in albinoes—the eye is unable to perceive objects placed in a very bright light. The development of this pigment in the eye is proportioned to its development in other portions of the body—hair, skin, &c.: hence, those who have fair hair have light blue eyes, and those with a swarthy skin have black eyes; also, those who have black eyes can tolerate a more glaring light than those who have eyes of a grey or azure tint.

The Retina.—This is the nervous layer of the eye (R R); it is the expansion of the optic nerve, and lies immediately within the choroid, which can be seen behind it because of the transparency of the latter. It consists of several layers of cells, fibres, granules, and rods, which are placed one over the other, and can only be seen with the assistance of the microscope. The retina is the "collodion plate" of the eye, upon which is photographed the multitude of objects, which are hourly and almost incessantly presented to the mind. Extending as it does over the whole inner surface of the ball as far almost as the iris, it constitutes a vast canvas for the reception of images, and, although extremely sensitive to *light** at every point but one, there is one locality in which impressions are received more distinctly than in others, and one place where there exists no sensation at all, and which is commonly designated "the blind spot." The portion of the retina which is capable of receiving the most delicate impressions is situate in the front-to-back axis, or rather in the hinder pole of this axis. It is the spot upon which the rays of light most usually impinge, and it has been said that to ensure perfect vision the image must be formed upon this region. The blind point lies at the entrance of the optic nerve. This latter enters the eyeball not in the centre of its back portion, but at a point nearer the inner surface; hence it is not often so placed, that images are thrown upon it. The following experiment demonstrates its existence. Holding the arms out at full length, bring the two thumbs upwards and together; then, having closed the left eye, regard the left thumb *fixedly* with the right eye, and while doing so slowly move outwards the right arm, and the right thumb will, in passing, become invisible

---

* I have italicised the word light, because some persons might suppose that it was also capable of feeling pain; but this is not the case.

HUMOURS OF THE EYE.

at one particular point, viz., as the image passes over the "*punctum cæcum.*" The retina ends in front by a toothed border, which dovetails with the choroid at a short distance from the union of the sclerotic and cornea.

The next question that suggests itself is, what is to be found within these coats? Looking upon the eye as a sort of capsule, we may for convenience of description divide it into two portions—back and front. These are filled with materials of different characters; those which are contained in the posterior part being solid, and those in front more or less fluid. The great bulk of the eye belongs to the hinder division, and comprises two distinct structures, which are perfectly transparent,—the Vitreous Humour, and Crystalline Lens. And first of the Vitreous Humour (v): this is a clear, solid, jelly-like mass, closely embraced behind and at its sides by the retina. It fills the ball from behind, to about the commencement of the cornea in front (where it is overlapped by the termination of the choroid), and is enclosed in a membranous bag which enters the humour, and divides it into numerous compartments.

The lens, or crystalline lens (c), is placed just in front of the vitreous humour, in a hollow, which the latter presents for its reception. It is a very small, lenticular body, double convex,—that is, elevated on both sides, and transparent; moreover it does not merely *lie* in the depression, but is fixed in its proper position by a circular band, which is attached on the one hand to the membrane which encloses the lens, and on the other to a sinewy fold, which passes around within the eye at the union of sclerotic and cornea. It is much flatter in front than behind, and is composed of about 200 layers of fibres, which are toothed at the edges, and adapted to each other beautifully. In front of the crystalline lens and the vitreous humour (which united fill up the whole of the back part of the eyeball), and behind the cornea,—intervening between the two, is a cavity, which during life encloses a fluid of a saline, transparent character—the Aqueous Humour (D). I have nothing more to say concerning this humour, save that it is almost divided into two portions by an exquisite circular partition, which has an aperture in its centre. This partition is named the Iris (I I), and in it lies the colouring matter which gives the peculiar shade either of brown, hazel, blue, or grey, to which much of the beauty of the eye is due. The small aperture is called the *Pupil*, and is recognised as a black spot in the centre of the eye, which increases in size in the shade, and diminishes under the influence of light.

How does this circular aperture get smaller or larger? By changes in the iris. The latter contains two distinct sets of muscular fibres; one series is disposed in concentric rings or circles, and when they contract, the orifice (pupil) is diminished. The second series is arranged in a radiating manner, in other words the fibres pass out from the central to the circumferential parts as spokes from the middle of a wheel, and when shortened during contraction they tend to widen or dilate the orifice, giving a very brilliant appearance to the eye.* Finally, the iris is lined at the back by a continuation of the choroid coat, and therefore it forms an opaque shutter for the sensitive surface of the retina. There is yet another structure which had almost escaped our notice,—this is ciliary muscle; it consists of a belt or zone of muscular tissue, which is attached to the cornea and sclerotic at their junction, and which passes round the whole eye in this region (L L). It is also united at its inner border to the crystalline lens, and *some* physiologists believe that it is the special muscle of this body.

The exact office of each of the mechanisms we have been studying will be explained in replying to the following questions :—

How do we see an object ?—I think the reader is now in a fair position to answer this question. The rays of light (*a g*, *b t*, fig. 78) which come from an object, strike upon the cornea (*d e*), and as it is transparent they travel through it, but in doing so they are bent from their former direction and caused to converge. Now, forming a cone, they pass together through the aperture of the iris (the pupil), impinge upon the crystalline lens (*n o p q*), and are transmitted, but are forced to converge still more, till they meet at a point (*r*) somewhere in the vitreous body. This is the focus, and from it they then *di*verge, and at last strike the retina; so that in this manner a picture (*u f o*) is thrown upon it. Strange to say, however, this picture is turned upside down, for, as the rays *di*verged after their *con*vergence, those which came from the top of the object passed downwards, and those from the bottom upwards.

How, then, do we see the object, as if the image on the retina was not inverted?—This question has been answered by many writers. One tells us it is certainly because of one circumstance; and another lays down the law with apparently equal force; but, somehow or other, a little consideration will invariably enable the reader to detect a blunder in the explanation. For myself, I am content to say I do not understand the reason, though I admit the fact; and that, till some one points out a more intelligible cause than those already put forward, I must think that the correction is due to some influence of the mind, which at present we are unable to comprehend.

Why cannot we at first see distinctly, when we encounter a bright light after having been for some time in the dark? And why, when we pass from the bright sunlight into a dark cellar, are we for a few moments confused?—I have given these two questions together, for their answers bear much upon each other. In the first instance, we have been in a dark chamber, and in order to see as clearly as possible, our pupils, or passages for the entrance of light, have been opened widely. On entering the well-lit room, the light enters too freely by the enlarged orifice; and so things are seen dimly, or we close our eyes for a few moments. This unpleasant sensation does not last long, for the iris, by contracting, diminishes the aperture,

* If a drop of a solution of atropia is placed upon the eyeball, a similar effect is produced in a very short time. This fact is so well known to many members of the fair sex that I merely allude to it without further comment.

regulating its size; and thus only the proper quantity of light is admitted. In answering the second question, I need only say that the pupil, which was diminished in size by the influence of the sun-light, was not sufficiently large on entering the cellar to admit light enough for objects to be seen at all.

That this proportioning of the size of the pupil to the brightness of the light does take place, the reader may demonstrate for himself by examining the size of the pupil (under different conditions) with the assistance of a small mirror. The iris, then, constitutes the shutter of the eye, and protects the sensitive retina from the baneful influence of excessive light. Its alternate contractions and relaxations are very good examples of what are termed reflex actions. Without any attempt at what a distinguished author has termed "*Bridgewater-writing*," I think we may well regard the iris as one of those innumerable evidences of the existence of that "high and mighty King of Kings" which biology offers to the mind of the reflective student.

Why are we at first unable to perceive a distant object, when for some time previously we have been regarding a near one? And why, after we have been looking at something in the distance, are we unable at first sight to perceive clearly any object which is placed near the eye? The first thing which these two questions suggest is, that the eye has the power of adapting itself to near and distant objects, so that each may be observed distinctly. If the eye remained unchanged as regarded the curves of its constituent lenses, the rays from distant objects would not be focused at the same point as those of near ones; consequently, either set would be perceived indistinctly. How the eye does change the form of its lenses, and what lenses are altered, are questions which have been answered over and over again, and whose answers [as given by physiologists] are of the most conflicting kind. Quite recently, a writer has asserted that the blood-vessels, by the pressure which they exert on the vitreous humour, cause this to be projected forwards "as a patient is on a water-bed," and therefore the lens to be advanced also; and that this change always takes place when we view a near object. Unfortunately for this theory (which, by the way, is not by any means a recent one), it is known,—*

Firstly,—That there is an aqueous humour between the lens and cornea, and that in order that these should be advanced it should be allowed to escape.

Secondly,—That in all cases of congestion of the eye there is not an absence of "distant vision;" and

Thirdly,—That the retina *might* not be able to sustain such severe pressure.

A glance at the construction of the eyeball will give the reader an insight into the true process of "adaptation," as it is called. Look at the position of the ciliary muscle—a structure which (as has been demonstrated in the eyes of birds) is undoubtedly contractile, and

* For my objections to this hypothesis consult the "British Medical Journal" for April 18th, 1863.

you will perceive that, situated as it is, a very slight contraction of its fibres must produce a constriction of that part of the "ball" at the junction of the sclerotic and cornea. Now, what will be the immediate result of this constriction? Simply, a bending inwards (all round) of the hinder border, or edge of the cornea. What will be the effect of this? An increased curvature of the cornea, and hence the capability on the part of the eye of perceiving near objects. Let us suppose that it is required to examine a very distant object. What happens then? The muscle relaxes, and the cornea, by virtue of its elasticity, which is very great, returns at once to its former position. That the iris has nothing to do with either near or distant vision has been shown in a case recorded by a great German surgeon. The patient had had his iris completely removed, yet his power of perceiving near and distant objects remained the same as it had been before the operation.

The above explanations lead us to another question,— *Why are some people short-sighted and others long-sighted?* In near-sighted people the cornea is extremely convex; hence its power of refraction is very great, and the rays are brought to a focus too soon to allow of distant objects being seen distinctly. If therefore we can by any means bring the rays coming from a distant body in such a manner that they shall strike the cornea more obliquely, they will not be brought to a focus so soon, but will form a clear image on the retina. Concave glasses produce this effect, and consequently these are the kind employed in the spectacles of near-sighted persons. In long-sighted people the cornea is too flat, and the rays are not brought to a focus soon enough—the image having a tendency to be formed behind the retina. This results from the rays not falling perpendicularly enough to be refracted sufficiently; and therefore we add to the ordinary refractile power of the eye by placing a convex lens in front of it. That long-sighted people have flat corneas, and short-sighted persons prominent ones, I think is evident to every one who possesses even ordinary observation.

Why are some objects one colour, and others another? I told you before: that white light is really composed of many distinct hues; and as we see all things by the aid of light which is reflected from them, it follows that if any body or object absorbs some of the coloured rays completely, only the remaining ones will be reflected to the eye and be recognized. Thus anything which absorbs all the red rays will appear green, an object will appear blue which retains the red and yellow rays, and so on.

It is extraordinary, that in some individuals the power of appreciating certain colours is not present. There are persons who cannot distinguish between blue and scarlet. This peculiarity is termed "colour-blindness," and is much more prevalent than it is thought to be by many.

It is stated that, on one occasion a tailor sewed together the different portions of a scarlet hunting-coat with blue silk, and presented it to his master with the utmost gravity, having been unable to detect his error, by reason of his inability to discriminate between the two colours, scarlet and blue. If we gaze fixedly and for some time upon a piece of scarlet paper, and then turn the eye upon some other object, it appears of a greenish hue. How is this? The retina having been exerted in staring at one colour, at length becomes insensible to *it*, and is capable of receiving only the two other colours (blue and yellow*), hence everything looks green.

This explanation enables us to understand the cause of another phenomenon, which is this: if after looking upon the sun we turn our eyes upon some other object, we observe a round black spot. In this case, that part of the retina which has received the direct rays has become *pro tempore* paralysed, and when we regard other objects, they are depicted on every portion of the recipient surface except the insensible round spot, which then appears as a black disk on the field of vision.

Having two eyes, we have simultaneously formed two **distinct** *images of every object: why then do we not see things* **double?** Because the mind associates the images, and when two of them are depicted on the retina, the mind by habit perceives but one. That the perception of a *single* object is in a great measure the result of habit, appears from the fact that when we press one of the eyeballs inwards, and so cause an image to fall upon a part of the retina, which has not acquired the property, a double picture is seen. In squinting also, where the rays are thrown upon a new portion of the retina, double vision results, but after some time objects are seen as usual.

Some physiologists tell us, that without the two eyes we could form no conception of solidity: this is not true. Close one eye, and with the other you gain nearly as perfect notions of the solidity of a box, a table or a bottle, as you did with the two. However, there is no doubt that we do get with the two eyes different views of the same object. Look, for example, at the back of a thin book which is held on a line with the nose, and at a short distance from it, first with one eye, and then with the other; and you will perceive that you have seen it from different points of view. In looking at it with both eyes, it is supposed that the mind summing up the two aspects has concluded that the object is solid, in the same way as a person who has seen two sides of one of the pyramids of Egypt forms an idea of its solidity, whilst had he been placed at once in front of but one side, he would have thought it but a gigantic wall. In the stereoscope, two pictures of a relief are placed beneath the eyes, and they are perceived as a single picture which "stands out" perfectly. This will not be the case if the pictures be *merely* similar ones; they must represent the images which would be formed upon the two retinæ when the object *itself* was viewed. In other words, they must be taken from different points of view; one must be the picture as it would be seen by the right eye, and the other such as would be projected upon the retina of the left.

* These are then termed "*complementary*" colours, as they are portions of the white light. Red, blue, and yellow are complementary to each other.

When a burning brand is **waved violently to** *and fro, why do we perceive a tape-like flame, in the form of* **a circle?** An image which has been formed upon the retina remains for about the one-tenth of a second, and after that vanishes. If, then, we can throw upon the sensitive surface a series of images of the same object within the one-twentieth of a second, it is evident that ere one has left, another will be formed, and so a continuous image will be presented. Now, if the images be thrown upon the retina so close to each other as to touch, while at the same time by the motion of the object they are caused to revolve in a circle, an appearance resembling a girdle of flame will be produced.

Our ideas of size, distance, surface, polish, and so forth, are evidently the result of comparison between the ideas derived from different senses. Thus, we feel the surface of a table, and acquire a sense of smoothness, and at the same time we observe by the eye that the surface reflects light well; we therefore form a notion of polish. In this instance we associate the two results, and ever afterwards the notion of one recalls the idea of the other, and when we see anything which reflects light well, we assume that its surface is smooth and polished. The same statement may be made concerning size and distance, as the reader will doubtless admit. The following remarks on this subject are of great interest:—

"A boy of four years old, upon whom the operation for *congenital* cataract had been very successfully performed, continued to find his way about his father's house, rather by *feeling* with his hands, as he had been formerly accustomed to, than by his newly-acquired sense of sight; being evidently perplexed rather than assisted by the sensation which he derived through this. But when learning a new locality he employed his sight, and evidently perceived the increase of facility which he derived from it. Among the many interesting particulars recorded of the youth upon whom Cheselden operated with equal success, it is mentioned that, although perfectly familiar with a *dog* and a *cat* by feeling them, and quite able to distinguish between them by his sight, it was long before he associated his *visual* with his *tactile* sensations, so as to be able to name either animal by sight alone. The question was put by Locke, whether a person born blind, who was able by his touch to distinguish a cube from a sphere, would, on suddenly obtaining his sight, be able to recognize each by the latter sense. The reply was given in the negative; and the experience of the cases just referred to, as well as of many others, fully justifies such an answer."[*]

From what has been already observed, we are justified in comparing the human eye to the camera obscura of an artist, and the retina to the sensitised plate which receives the image that is, as it were, "developed" by the influence of the mind. How perfect an apparatus it seems when we reflect on the fact that the pictures which it photographs represent often areas of square miles, and are only to be numbered by billions!

---

[*] Manual of Physiology, by W. B. Carpenter, M.D., F.R.S., F.G.S., &c. 3rd Edition, p. 629.

EYELIDS—TEARS—DISEASES OF THE EYE.

The eyelids are fleshy membranes stretched over a framework of gristle. They are covered outside by a layer of the skin, and within by a delicate fold of the exquisite mucous structure which wraps the eye. By virtue of the fleshy tissue which they contain they can be brought together or separated at will; but we cannot keep them apart for any considerable time; firstly, because the muscle fibres which sustain the upper ones become exhausted for a moment, when the lids fall; and secondly, because the surface of the eyes becomes dry and painful, owing to exposure to the air. The closure of the lids spreads a liquid secretion over the ball, which lubricates and protects it from injury. The source of this secretion I shall state presently. The eye-lashes serve a double purpose: they shade the eye, constituting a kind of **Venetian** blind, and they prevent the perspiration of the forehead touching upon the surface of the cornea, and producing irritation. Of course this second office is in these climates a sinecure. Not so, however, in tropical countries, among the natives of which we find the eye-lashes and eye-brows fully developed. The function of the eye-brows is almost self-evident. They prevent particles of dust and drops of liquid from passing beyond their margins.

The tears are formed from the blood in the same manner as the saliva and other secretions, by the action of two small glands called *lachrymal*. In structure each of these resembles the salivary glands, which have been described under the head of digestion, and is placed in the outer part of the bony socket (orbit) of the eye. It is somewhat of the form of an almond, and pours its secretion upon the eye through six or eight little tubes. As soon as the tear-fluid has reached the margin of the lids, it passes from them over the surface of the cornea, till it reaches the inner corner of each eye. Here it flows through two small openings into a little reservoir, which in its turn communicates with the nostrils. This arrangement explains two phenomena—the passage of tears over the face, and through the nose—the first being due to an overflow of the cistern, the second to the connection between the inner part of the nose and the orbit. Besides the structure above enumerated, each lid is provided with from twenty to thirty little tree-like glands, whose apertures are placed at the margins, from which issues an oily secretion. The office of this is to prevent the lids adhering to each other, and also to form a temporary gutter for the tears. It is it also which, when the lids have been approximated for some time (as in sleep), develops the ugly gluey material found uniting the lids on rising in the morning.

Before closing this chapter, let me give very popular definitions of two common forms of disease of the eye:—

First, Cataract.—In this disease the crystalline lens becomes opaque, so that rays of light do not travel through it sufficiently to form an image. It is cured by cutting open the cornea, and extracting the lens.

Second, Amaurosis.—In this disease the retina loses its power of transmitting to the mind the images which are formed on it: it is, in fact, paralysed, and cannot be cured by any *surgical* operation, though the other structures are perfect.

CHAPTER XV.

The Nose—Perfumes—Power of Scent—The Tongue—Lingual Papillæ—
Taste—The Ear—Sense of Hearing—Nature of Sound—Touch.

THE sense of smell is seated in the nostrils (*a b c d e*, fig. 80), and these are certain cavities contained in that organ familiarly known as the Nose. The nose is a kind of irregularly-shaped box with four openings—two outer ones above the upper lip, and two inner, placed in the expansion of the gullet, called *pharynx*. It is, then, one

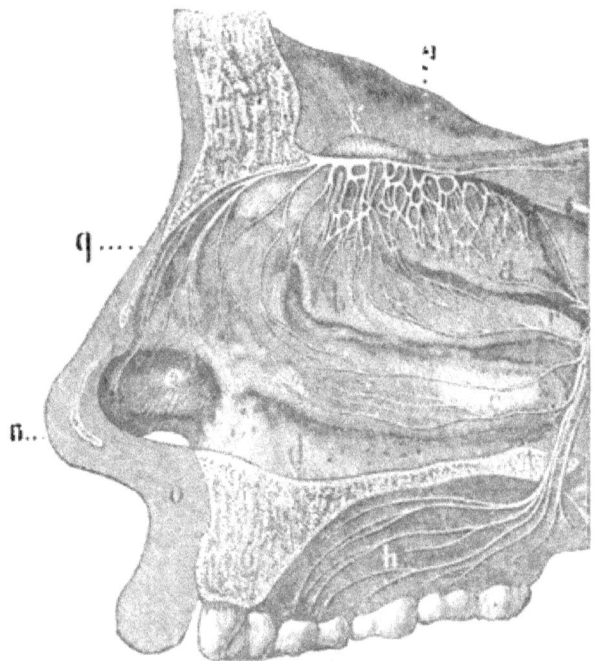

Fig. 80.—*a b c d r s*, membrane of nostrils; *e*, wing of nose; *n*, nose; *q*, nose-bones; *o*, upper lip; *g*, jaw-bone; *h*, hard palate; *m*, bone of skull; *k*, olfactory bulb; *p*, branches of the latter.

of the channels through which air passes to and from the lungs, and when the mouth is closed, respiration may be satisfactorily carried on with the assistance of the nostrils.

The nose is divided within by a vertical partition into two distinct compartments, and each of these is further divided (partially) by transverse walls of bony tissue which spring from its sides. The entire surface of these osseous walls is clothed by a soft, velvety mucous membrane, bountifully supplied with blood-vessels, and having numerous small glands embedded in its substance. To it are also supplied hosts of filaments of nervous tissue, which stream down from the "olfactory bulbs,"* and are the organs through which the impressions of odours are conveyed to the cerebrum. There are other nerves which are distributed to the nasal membrane, conferring upon it *common* sensibility, the power of secreting mucus, etc., which we shall not consider in so general a review of the apparatus as the present. The nostril being placed as a sentinel upon the entrance to the lungs, and having the power of examining everything which is about to enter in the gaseous form, necessarily constitutes a very important means of protecting these organs from injuries to which they would otherwise be constantly liable. For example, pure air, in traversing the nose-passages, produces no unpleasant effect; but let it be adulterated to a small extent with a foul vapour, such as sulphuretted hydrogen, and the result is, that the noxious gas being perceived by the sensitive membrane of the nostrils, a conception of danger is formed in the mind, we immediately refuse to expire such an atmosphere, and rushing from it in a state of intense alarm, seek fresher and purer air in some other locality. The nose is, therefore, in some measure, the guard-house of the lungs.

That it is of use as a special organ by which we have the power of gratifying our sensual desires, is evident to those who love the perfume of flowers. It also serves another, and more utilitarian purpose, though rather in the lower animals than in man,—the power of recognizing certain races by the odours which emanate from them. Through this faculty the beasts of prey are enabled to "scent" their food, and the Peruvian and other Indians can distinguish between various tribes, even though the sense of sight be absent.

Just as the eye becomes dead to a certain colour when it has been gazing on it for some time, so may the nose become insensible to any odour to which it has been long exposed: thus surgeons, anatomists, and apothecaries become accustomed to the (to others) very unpleasant smells by which they are surrounded, so that after a while they do not perceive them. The likes and dislikes of persons relative to odours are very remarkable: to one the smell of asafœtida is delicious, to others disgusting. Some hysterical women are delighted by the smell of burnt feathers, an aroma which is productive of vomiting in those who enjoy good health. Substances alone which give off a part of their component atoms in the form of vapour, are recognized by the nose. From this we conclude that the sensation produced in the nostrils† is due to the lodgment of an infinitesimally minute particle in the mucous membrane of the nose. That an exceedingly small particle is sufficient, has been shown by experiment. If

---

* Two root-like grey projections, which originate in the front portion of the brain, and from their position on the base of the skull above the nose, send down clusters of fibres to the membrane of the nostrils.

† Or rather referred to the nostrils.

we make with spirit an extract of *one* grain of musk, and add to it *thirteen million times* as much water as there was of musk, then each of these thirteen million measures of water will contain the one-thirteen-millionth of a grain of the odorous substance. And as each of these measures has a distinct perfume, it is evident that we can detect the one-thirteen-millionth of a grain of musk by the aid of the nostrils.

A grain of musk which had been exposed in a room for *ten years*, and which during that period had given a perceptible perfume to everything in its vicinity, was found afterwards to have lost no weight appreciable by the most accurate balance. In smelling it is necessary that the mucous membrane be moist, for it has been found that when the nerve which endows the nostrils with the power to secrete fluid is cut, the sense of smell is lost; and furthermore, that when glycerine is rubbed over the surface with a feather the sense is regained. If there be an excessive secretion of moisture, as in influenza, the power of perceiving odours by the nose does not exist.

It is yet a question, whether the "olfactory" sense is resident in the so-called olfactory bulbs, and a very interesting case, bearing upon the question, has been recorded by a learned French physiologist. When dissecting the head of a female subject he found that both "olfactory lobes" and the nerves of smell were absent, and could never have been present. This led him to make inquiries among the friends of the deceased, the result of which was a general statement on their parts, to the effect, that when living the woman's sense of smell had been very perfect, and that pleasant and unpleasant odours affected her as they did other persons; thus proving that without what are thought to be the essential portions of this sense-organ the woman enjoyed the faculty nevertheless.

Organ of Taste.—The faculty of taste lies in the Tongue, whose function is a threefold one,—

      1st. It assists in swallowing.
      2nd. Speech is in great part effected by it.
      3rd. Taste or gustativeness is seated in it.

Look down upon the upper surface of the human tongue (fig. 81), and you will perceive a great number of little elevations, which anatomists call *papillæ*. Besides those which you see there are many others, which are concealed from your view. There are three sorts of perceptible papillæ: — 1st. *Conical* forms (c), which project considerably from the surface, and are scattered here and there over the point and borders of the tongue. 2nd. Whip-like varieties (k), springing from the general surface of the organ, and surrounding the conical ones which lie in their midst. 3rd. *Circumvallate* ($h\ l$), or ditched-round papillæ. These, like the others, are projections, but they have peculiar features which distinguish them from the rest. Each elevation is surrounded by a little trench, which is fenced in by an outer wall of membrane—they are miniature towers encompassed by their moats. The circumvallate papillæ are not, like the other sorts, distributed over

THE TONGUE. 145

the surface, but are confined to one locality,—the base or root of the tongue (see fig. 81). All the papillæ are composed of the following structures:—

(a) A complex knot of very small blood-vessels.
(b) Within the knot a beautiful leash of delicate nerve-fibres, which receive the impressions.
(c) A covering of transparent, almost structureless membrane.

The tongue is moved by a complicated machinery of muscles, which make up its substance; and having three properties—those of con-

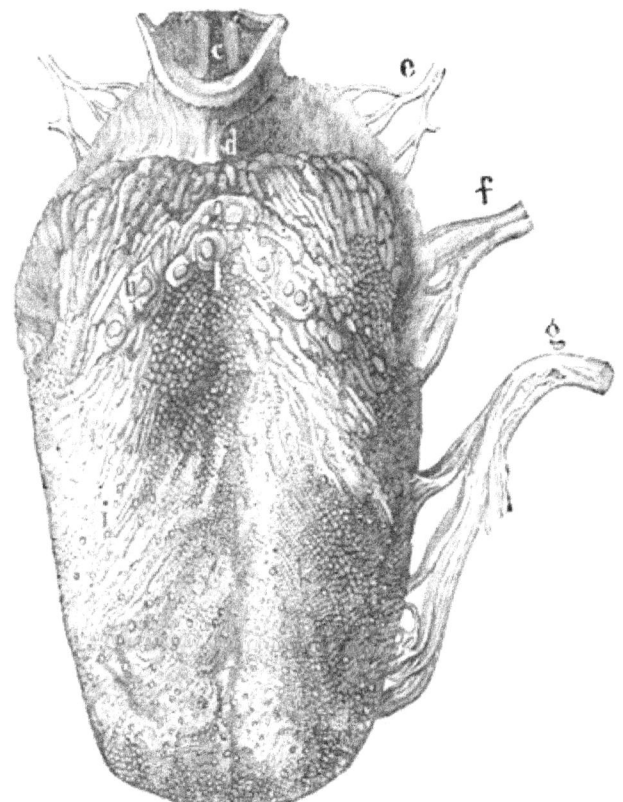

Fig. 81.—The Human Tongue.

tractility (motion), common sensibility, and sensibility to taste—it is supplied, as we might have expected, with nerves from three different sources (*e f g*, fig. 81). We found that the organ of smell was only sen-
L

sible to substances in a state of vapour; on the other hand, the organ of taste is only sensible (in a gustative sense) to substances in the liquid condition. Such materials as arsenic, alumina, etc., give rise to no taste,—are tasteless, because they are insoluble in the liquids of the mouth. We are accustomed to regard the tongue as the exclusive organ of taste, and as having the power of receiving gustative impressions at any point on its surface. Both these notions are erroneous.

Place a small piece of salt, or a bit of aloes, upon the front of the tongue, and so long as the substance does not pass to the root, the peculiar flavour will not be detected. The sense is absent in the following parts:—

The front half of the tongue on its upper surface,
The gums,
The inner surface of the cheeks,
The skin of the hard palate;

and is present in the arch of the palate beneath which the food passes, the soft palate and its projection the uvula, and in the membrane which attaches the tongue to the trap-door of the organs of voice; besides being in its most highly-developed condition in the *circumvallate* papillæ which stud the surface of the base (or root) of the tongue.

The sense of taste, like that of smell, is often very delicate. Bitter and acid substances are detected when present in smaller quantities than sweets; thus, the taste peculiar to each can be perceived with the *one-six-thousandth* part of a grain of oil of vitriol, the *one-three-thousandth* of a grain of extract of aloes, and the *one-four-thousandth* of a grain of sulphate of quinine.

When a sick person complains of a nauseous taste, it is probable that the altered blood has thrown out a peculiar secretion to which the unpleasant flavour is due. There has been much controversy as to the exact function of each of the nerves which supply the tongue, a great many conflicting statements have been made, and the difficulty in drawing a rational deduction is so considerable that we shall do well to leave the question "*sub judice;*" for its complexities would serve rather to retard the *general* reader's progress than to assist it.

Organ of Hearing.—This, as everybody knows, is the Ear; but though all are familiar with its external features, but few are aware how interesting an apparatus it is within. The ear is composed of two parts and a connecting channel—the outer ear, inner, and auditory passage (fig. 82). The external ear is a gristly, skinny structure, adapted to catch the various vibrations of the atmosphere which are produced by bodies that emit sounds. The central cavity, or funnel, which is continuous with the passage, is called the *concha;* and the other portions are known to anatomists by numerous designations, which I need not here allude to.

The inner mechanism is lodged in the substance of one of the bones of the skull, and is supplied with a nerve—the auditory (n, fig. 82)—presiding over the sense of hearing especially. It does not com-

ANATOMY OF THE EAR. 147

municate directly with the external ear, for, at the end of the passage is fixed more or less tensely a membranous cover, which is termed the *tympanic* membrane—or, more frequently, the *drum*.

This tympanic membrane is a kind of partition between the two divisions of the organ. It does not, however, form the immediately

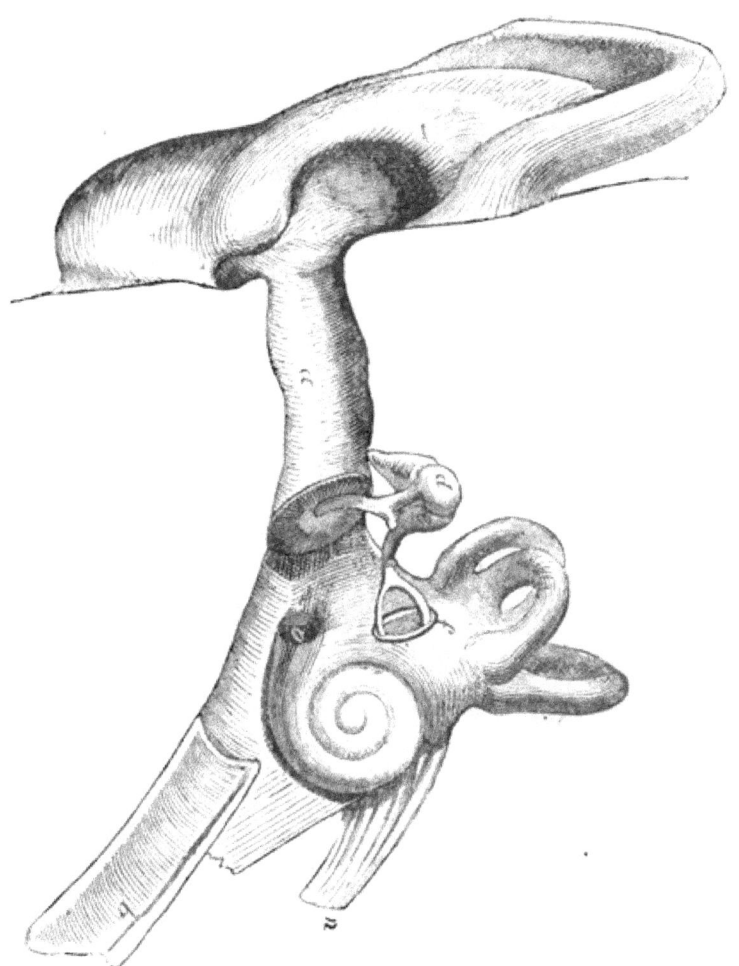

Fig. 82.—Inner and Outer Ear.

outer boundary of the inner mechanisms—a cavity lying between these latter and the drum styled the middle ear. How then is the sound-vibration conveyed from the tympanum to the internal apparatus where the nerve which is to receive the impression lies? By an

148   POPULAR PHYSIOLOGY.

exquisitely small chain of bones,* which are placed end to end, one of whose extremities is fitted to the drum and the other attached to the true ear. Of what does the real auditory organ consist? Of three portions, collectively called the labyrinth—a name sufficiently indicative of their complex arrangement and character. These three portions are—

> 1st. The *semicircular canals*.
> 2nd. The *snail-shell*, or *cochlea*.
> 3rd. The *vestibule*, a cavity intermediately placed.

The labyrinth is provided with two apertures, covered during life with little membranous shutters. One is called the round (*o*), and the other the oval opening (*f*), and into the latter is fixed the broad end of the *stirrup* bone. Under ordinary circumstances this labyrinth is filled with fluid,† and by this the impressions conveyed by the chain of bones from the drum to the *oval opening* are carried to the different terminations of the nerves of hearing.

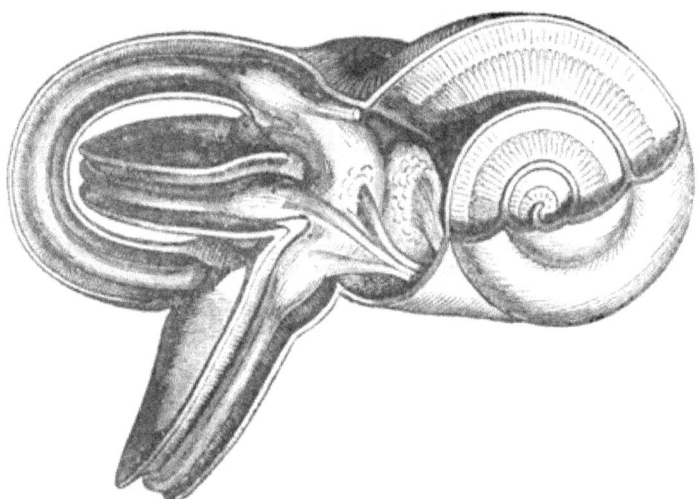

Fig. 83.—Inner Ear exposed.

The semicircular canals (fig. 83 left side) are three bony tubes, curved in a somewhat semicircular manner, and arranged at right angles with each other. These dilate at their extremities, and in the expansions are placed the swollen ends of branches of the nerve of hearing. The function of these peculiar structures has not yet been ascertained with certainty, but from what is known of them it seems not unlikely that it is through them that we are enabled to judge of the locality from which a sound has travelled.

* These on account of their shape have been named (from without inwards) the *hammer* (*d*), the *anvil* (*e*), and the *stirrup* (*f*). (See fig. 82.)
† It is yet to be *proved* that there is no fluid in the snail-shell or cochlea.

HOW WE HEAR SOUNDS.

The snail-shell or cochlea (fig. 83, right side) has an outline which its name hints at. It is a canal coiled round a central axis for two and a half turns, and it communicates with the vestibule—which is simply the space intervening between this structure and the canal—by an aperture. Its spiral cavity is divided by a partition into two channels, and on this partition is placed a beautifully toothed membrane, whose office is not well understood. The tube composing the cochlea expands into a dome-shaped chamber at the top of the spire, and has supplied to it an extensive branch of the auditory nerve. From several observations regarding the development of this mechanism, and the powers of animals to appreciate the *pitch* of sound-impressions, it has been concluded as probable, that its use is to enable us to acquire an idea of the **pitch** (or length of the scale) of different sounds.

You may perhaps ask what is the middle ear filled with? **It contains** only air. But how comes air in such a position? By a very beautiful contrivance. There is a half-bony, half-membranous canal* (*b*, fig. 82) which passes downwards from the cavity, and opens **into** the back part of the throat, and which in health allows the **air** to travel upwards or downwards, according to circumstances. **If** the air in the middle ear expanded, it would (had it no means of egress) cause the drum to bulge out, and so would seriously interfere with the reception of sounds. Of this the reader may convince himself by closing the mouth and nostrils and attempting to breath *out*; he will then feel the drum pushed gradually outwards,—an intense singing in the ear accompanying this alteration. The connection through this canal of the middle ear and throat explains how it is that an attack of cold in the latter affects the ear **also**, giving rise to a sensation of "*dinning*," and to partial deafness.

Having examined the three divisions of the **ear, we can** now undertake the consideration of the way in which the **sounds** (or rather the vibrations; for sound is the name **of an** idea) are propagated from **the drum to** the terminations of the nerve of sense.

Glance at fig. 82, and on it **map** out the route **of** the vibrations. I strike **the** string of a violin and instantly it is thrown into **a** flutter, **moving** from side to side in such a rapid manner that the movements **are** hardly perceptible; at the same time a sensation of sound is communicated to the mind. How does all this take place? When **the** string was touched, owing to its tenseness, it commenced a series of vibrations—of imperceptible movements from side to side—striking the air in its vicinity. By this the more distant air **was** caused to move, until at length that portion **of** the atmosphere **in the** neighbourhood of the ear was affected, and finally the drum of **the** ear itself. The column of air which was thus caused to undu**late is** called a wave; the wave began at the violin string, and tra**velled** onward at the rate of about 1,125 feet in the second, till it **ended at** the drum or membrane, from which we shall now trace its **progress** inwards.

The drum having received the **vibration which has**, so to speak,

---

* The Eustachian tube.

traversed the auditory channel (*a*), transmits it to the hammer-bone (*d*); this gives it to the anvil (*e*), from which it is forwarded to the stirrup (*f*), and at length arrives at the *oval opening* of the internal ear. Arrived here, it (the vibration) is communicated to the fluid which pervades the labyrinth, and is *possibly* conveyed in part through the semicircular canals and the snail-shell (fig. 83). In both of these organs it strikes against the recipient terminations of the nerve—in the first it gives rise to a notion* not only of sound, but also of the direction in which the sound travelled (?); whilst in the second an impression is produced which gives an idea of the *pitch* of the sound emitted by the string. I said that the vibrations into which the string was thrown were communicated to the air, and from it to the ear-drum; from which you will draw the inference that if there was no air no sound would be heard; and this is really the case.

I place a bell in the upper portion of a glass shade in such a position that when the shade is moved the bell will be caused to sound; if now I move the shade you will hear the sound of the bell distinctly. In the next instance I fix the shade over an air-pump, and having worked the instrument for a few moments, then ring the bell, and mark at once that the sound is fainter than before. I continue to work at the pump till I have thoroughly exhausted the air; now I again ring the bell, but though I can distinctly perceive the tongue striking against the sides, the bell is dumb, the sound is no longer distinguished. I again admit the air, and all goes on as before.

From the above experiment you are led to believe that the air conducts the vibrations to our ears, but you must not conclude that in *air alone* resides the power; wood, iron, and water can also convey sound very perfectly. The newest form of stethoscope† is composed of a solid rod of whalebone or cane, and the sounds of the heart can be detected quite as distinctly with *it* as with the older form of the instrument. The latter was hollow in the interior, and impressed the patient with such an idea of fire-arms, that it was a very formidable weapon in the hands of those physicians who, not possessed of much originality of their own, delighted to ape the Abernethy.

The physiology of the organ of hearing is even yet involved in so much uncertainty, that to enter upon loose and unscientific explanations of well-known phenomena would do no service to the reader, who (if he has carefully perused the foregoing paragraphs) will be in possession of as much information upon the subject as, for the present, he can hope to obtain.

Our ideas of the distance of sound are derived from the intensity of the vibrations, and from habit. That this is true **is** evident from the fact, that if a sound be produced near us, but yet of the same intensity as it would have had had it come from a distance, we will

---

* **Co-operating, of** course, with the brain and mind.

† **The credit of** this discovery is, I believe, due to Dr. Corrigan, of Dublin.

be deceived as to the locality from which it emanates. This, in fact, is what is done by all those who possess the power of ventriloquising.

It is stated that sounds of a very small number of vibrations per second (14 to 16), are unappreciated by the human ear, and that those which are composed of more than 64,000 in the second are also imperceptible. It is, however, quite sufficient to excite our wonder to know, not only that notes within these far distant limits are cognizable distinctly, but that during a single second a string may be caused to vibrate *sixty-four thousand* times.

The faculty of hearing may be developed in different directions: one man may possess great acuteness as regards the perception of sounds generally, another as regards the perception of musical sounds, and so on; but it is unquestionable that the auditory sense may be much increased by cultivation. To quote the language of one of the most philosophic writers in the English scientific world: "The watchful North American Indian recognizes footsteps, and can even distinguish between the tread of friends and foes; whilst his white companion who has lived among the busy hum of cities is unconscious of the slightest sound. Yet the latter may be a musician capable of distinguishing the tones of all the different instruments of a large orchestra, of following any one of them through the part which it performs, and of detecting the least discord in the blended effects of the whole—effects which would be, to the unsophisticated Indian, but an indistinct mass of sound. In the same manner, a person who has lived much in the country is able to distinguish the note of every species of bird that lends its voice to the general chorus of nature; whilst the inhabitant of a town hears only a confused assemblage of shrill sounds, which may impart to him a disagreeable rather than a pleasurable sensation."

Sense of Touch.—When we bring the ends of our fingers into contact with any object, an impression is immediately produced upon the surface of the skin, which is applied to the object, and this impression being conveyed by the *sensitive* nerves to the spinal chord and brain, an idea concerning the surface of the thing felt is forthwith formed by the mind. We say we have *felt* it. It is soft or hard, or smooth or rough according to circumstances. The power which we thus possess of conceiving of the nature of surface, without the aid of the eye, is called the faculty of touch. The faculty is seated in the skin generally, but particularly in the skin of the fingers. It is not, however possessed by the skin, but by certain structures in it, which communicate with the terminations of the truly sensitive nerves. To be plain, it resides in the extremities of these nerves.

How do we prove that the sense *is* resident in the nerves? Easily enough. We walk into one of the wards of a hospital, and sitting beside the bed of a patient who is suffering from such a disease of the spinal chord that the latter is virtually amputated, we with a feather tickle the soles of his feet. This produces an exceedingly unpleasant sensation in a healthy man, but in the present case it is *unfelt*. Nextly, we place our hands upon his limbs, and he tell us that he does not feel them, and that had he not seen their position he would not have believed that they were in contact with him. Why is the sense absent here? Because the spinal chord being divided, the line of communication between the surface of the limbs and the brain is severed—the telegraphic wire has been cut, and the messages to the mind are no longer transmitted. When a portion of the skin of the finger (cut at right angles to the surface) is placed in the field of the microscope, the observer may perceive beneath the outer or scarf-skin, a vast number of little teat-like elevations, called *papillæ* (fig. 84). To each of these passes a small blood-vessel, and also a minute twig from the sensitive nerve: the first forms a few loops round the papillæ; the second enters its substance.

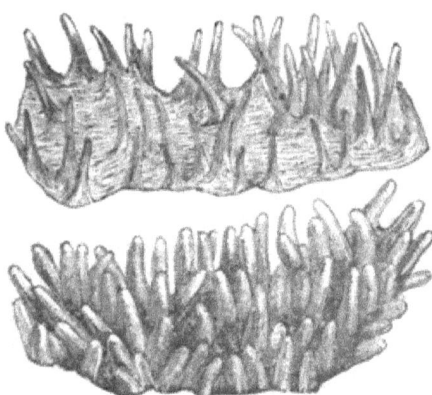

Fig. 84.

Anatomists have also discovered a microscopic body in the interior of many of these papillæ. It is in form like a fir-cone, and according to some is hollow in the interior—to others, is a solid structure. At this body the filament of the sensitive nerve ends, either by entering its substance and passing through its centre, or by winding round it in a spiral manner (*b c d e*, fig. 85).

As these little organs have not been invariably found co-existing with the tactile sense, it follows that they are not essential to it. That they may nevertheless intensify the faculty of touch when they are present, seems, as has been suggested, not unlikely.

The number of the elevations I have described is indeed astounding; they (papillæ) are placed in double rows, containing about sixty each; and as there are no less than forty of these bi-linear series in the square inch, it results that within this small space there are no less than *two thousand four hundred* distinct points, each capable of receiving some impression from an external object. This statement is true only for the skin of the fingers, where the elevations are most abundant; in other localities the papillæ are much more sparingly distributed. For this reason, the sense of touch is more highly developed in the

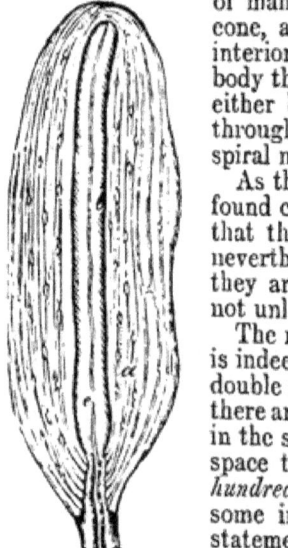

Fig. 85.—A Pacinian Corpuscle.

hand than in other portions of the body. A German physiologist proved this very ingeniously. He took a pair of compasses, and having brought the legs close to each other, he placed them in contact with the skin of the hand, and found that a sensation was produced, such as would have resulted from the application of two different objects; from this he concluded that the two points were brought into contact with two distinct papillæ. Next, he touched a portion of the skin of the back (keeping the legs of the instrument at the same distance from each other as before), and discovered that but one point was perceived, because there were not two sensitive elevations sufficiently close to be in contact with both ends of the compass. He now increased the distance between the legs, and on again applying it to the skin, remarked that both points were perceived. By an extensive series of experiments of this description, he was enabled to arrive at the following conclusions:—

That the two points can be perceived and distinguished { When only half a line apart, by the point of the tongue.

,,   ,,   { When about five lines apart, by the skin of the finger.

,,   ,,   { When fourteen lines apart, by the skin of the back of the hand.

,,   ,,   { When **thirty lines apart**, by the **skin of the back.**

The skin may be regarded as possessed of two senses,—one which enables us to form an idea of the shape and surface of objects, the other, which is *perhaps* an intensified condition of the first, which gives rise to a conception of pain. The latter is common to the flesh and skin, but the former belongs solely to the surface of the integument.

Besides these properties of the tactile papillæ, the skin enables us to form an estimate of temperature. This estimate, however, is in great part a relative one, and depends upon the actual heat of the skin or part of the skin applied to the object. If we dip the hand into water at 104° and afterwards into a fluid at 89°, the latter will appear cold; but if the hand be first placed in water at 68°, then that at 89° will appear to be lukewarm. A person may dip his hand into a crucible of molten lead, and if it be allowed to remain for but a moment no injury will be sustained. This is because a cloak of vapour is formed round the surface of the hand, owing to the evaporation of the fluids of the skin, and this layer of moisture protects the hand from being burnt. Lecturers on physics are accustomed to perform an experiment whose principle is the same. They make a platinum spoon *red*-hot, and pour a drop of water into it; the water is not boiled, but assumes a spherical form, and remains upon the spoon. This is because, at the moment it was placed upon the red-hot surface, the outer layer of the fluid was converted into vapour, which enveloped the globule, and prevented the action of the surrounding temperature upon it.

It is strange that when a large surface of the skin is exposed to the influence of warm or cold water, the fluid appears of a higher or lower temperature than if a small portion, as the finger for example, were dipped into it.

Pain, such as that experienced in case of burns, can only be produced by substances whose temperature is less than 50°, or over 122°.

Our conceptions of weight are in some measure dependent upon the portion of the skin to which the bodies are applied; thus, if a billiard-ball be allowed to roll down the face, from the cheek-bone to the lips, it will appear to increase in weight as it approaches the latter, because the sense of touch is more highly developed in this locality than in the upper portions of the face. In all cases of the measurement of weight by the hands, we are likely to be deceived, unless we bring in the other senses to our aid. For instance, we see the object, and from its size form a vague notion of its heaviness. Then by touch we know whether it be porous or dense, a good conductor of heat or not, and we draw very important conclusions from these observations. Finally, we lift it, and by a mental comparison between the amount of muscular exertion required to raise *it*, and necessary for the elevation of some other body whose weight is known, we decide how heavy it is. Habit and experience have much to do with the development of this sense. We fancy when blindfold that the same body placed in each hand is heavier in the left than in the right. Preconceived ideas influence our decisions to a great extent.

When Sir H. Davy discovered the metal potassium, which is so light that it will float on water, and placed it in the hands of a friend, the latter remarked, "Bless me, how heavy it is."

I bandage your eyes, and taking your fore-finger in my hands, I place it on the table, and press it against the surface, first lightly, then more heavily, and you exclaim, "You are moving my finger over some globular body." I reverse the order of pressure, and you say, "My finger now passes over some hollow substance." The correcting sense (sight) is absent, and the organ of touch being excited in the same manner as it would have been in passing over a convex and concave surface, you really conceived that in each case the plain surface of the table had one of these two characters. The tactile sense is usually more perfect among the blind than in ordinary individuals. It has been said of a celebrated Cambridge professor that, though blind, he could with the utmost facility distinguish between genuine and spurious medals by the sense of touch alone; and instances have been recorded of blind men who possessed the faculty of discriminating by touch alone between textures of different colours. In order to show the deceptive character of the sensations arising from tactile impressions, schoolboys perform the following experiment:—*Cross the fore and middle fingers of a person whose eyes are closed, and place a marble in such a position that it touches the extremities of both fingers. The person will receive an impression the same as that which would have resulted from the actual contact of two marbles* (fig. 86). What is the cause of this deception? When the marble is placed simply between two fingers there is no deceptive impression, because these are felt by the surfaces of the skin, one convex outwards and the other convex inwards; and these two we unite by one judgment to the conception of a solid ball. In the exceptional and deceptive instance, the two convex surfaces appear (on account of our habitual conceptions) to be both in the same direction, and we form the idea that two little spheres are present. This is the usual explanation, and we take it, as people say, "with a grain of salt." I do not commit myself to it, merely regarding it as an agreeable solution of the difficulty till some better one turns up.

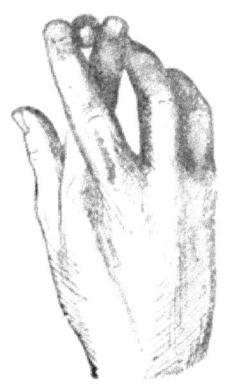

Fig. 86.

If I were to pile up experiments and observations of the kind I have just glanced at, I should soon have a "monster pyramid" indeed—a huge curiosity shop of ill-ascertained facts, useless statements, and miraculous phenomena, equally puzzling to the reader and myself. But why inflict upon society a host of threadbare anecdotes about this man, and that, and the other? Enumerate phenomena bordering upon the marvellous—*cui bono?* Let the accumulation I have exhibited suffice, and should the reader desire, let him peruse the works of those modern writers who revel in anecdotes, and love to indulge the public with *sensation* science, for statements and explanations of a more exciting character.

It is customary to introduce into works on physiology some description of the development of the human body, to discuss the question how from an almost microscopic germ the complex organism, containing muscles and veins, skin and bone, is produced; to retrace the various footprints upon that ill-defined intricate path, along which the being has journeyed to its complete from its immature condition. I admit that there are few points upon which attention could be concentrated with more interest or advantage; but, at the same time, I am fully aware that, from the character of the remarks which must necessarily be made to render the subject intelligible, and from the associations which the requisite allusions would call up in the minds of some, such a question could not be dealt with in a work intended, as the present is, for the improvement of the physiologically-ignorant of both sexes.

CHAPTER XVI.

CONCLUSION.

We have now arrived at the terminus of our physiological railway. We have passed slowly, it is true, along a lovely country, and it behoves us, now that the carriage doors are closed, and we are seated in our homes, to review mentally the ground over which we have travelled. It is to be hoped that we have not undertaken the journey in the spirit of mere conventionality, that we have not been conveyed along the line with closed eyes, and that our final object has not been the wish to be able to say we have been to science-land. If we would travel for the sake of having to say we *have* travelled, we should recall the lines of Hood:—

> "Some minds improve by travel, others rather
> Resemble copper wire, or brass, which gets the narrower by going farther."

No, dear reader! I trust our eyes have been open to the wonders of the science we have so far devoted ourselves to. We have not forgotten the maxim that "knowledge is power;" and what knowledge is superior to that which physiology affords to the earnest, truth-seeking student? What, then, is the fruit of our labour? How have we been benefited by the study we embraced? We have learned something of all those complicated phenomena which are commonly summed together in the term Life. It is not to be supposed that by life we understand a force of a *special* nature operating upon living bodies, nor a combination of those ordinary forces which regulate the action of inanimate matter; but rather that it is the operation of the *one* power which pervades the universe, and which, under different circumstances, exhibits itself in the production of different phenomena. We have observed how the two great classes of living beings are mutually dependent upon each other; how the various processes of digestion, absorption, circulation, and respiration, are merely the results of ordinary physical agencies, and that no peculiar *vital principle* need be invoked to explain the mode in which they are achieved; and how the special senses, sight, hearing, touch, taste, and smell, are to no slight or trifling extent under the control of the ordinary laws of physics. We have surveyed, and occasionally with scrutinizing care, the minute structure of what to the naked eye is almost "without form and void;" and by the assistance of the microscope have brought even new worlds within our ken. Having for our motto, "Know thyself," we have examined the most elaborate work of the Creator, the highest summit in the grand mountain-range of organization—" God's own image." Assuredly, the contemplation of our *very selves* has exalted and expanded our minds, and knowing, as we do, that "the proper study of mankind is man," we can each exclaim with the divine writer, "O God, I am fearfully and wonderfully made."

A caviller might insinuate that the views I have put forward are materialistic in tendency. I cannot think that materialism has been advocated, even where principleist doctrines have been objected to. It is not because we endeavour to account for the actions of living bodies by referring to the ordinary laws of physics, that we are, therefore, guilty of materialism; for, so long as we regard the properties of matter as having been originally impressed upon it by the Divine will, so long must we regard their continuance as dependent upon the same Almighty fiat which called matter itself into existence "in the beginning." In the words of one of England's most philosophic bards :—

> "All are but parts of one stupendous whole,
> Whose body nature is, and God the soul ;
> That changed through all, and yet in all the same,
> Great in the earth as in th' ethereal frame ;
> Warms in the sun, refreshes in the breeze,
> Glows in the stars, and blossoms in the trees,
> Lives through all life, extends through all extent,
> Spreads undivided, operates unspent,
> Breathes in our soul, informs our mortal part,
> As full, as perfect, in a hair, as heart ;
> As full, as perfect in vile man that mourns,
> As the rapt seraph that adores and burns :
> To Him no high, no low, no great, no small ;
> He fills, He bounds, connects, and equals all."

LIST OF WORKS,

Which may be consulted by those who are desirous of pursuing the subject of this Work further.

Carpenter, W. B., M.D., F.R.S.—"Principles of Human Physiology."

"Manual of Physiology."

Dalton, John, M.D., &c. — "A Treatise on Human Physiology, designed for the use of Students and Practitioners," with illustrations. 8vo. Philadelphia. 1861.

Kirkes, Wm. Senhouse, M.D., &c.—"Handbook of Physiology," 8vo.

Knox, Robert, M.D.—"Man—His Physiology."

Lewes, G. H.—"The Physiology of Common Life." 2 vols.

Müller, Johannes.—"Elements of Physiology." Translated by Dr. Baly.

Todd, R. B., M.D., and Bowman, Wm., F.R.S.—"The Physiological Anatomy and Physiology of Man." 2 vols.

Valentin.—"A Text-book of Physiology." Translated by Dr. Brinton.

Wagner, Dr. Rudolph.—"Handbook of Physiology."

"The Cyclopædia of Anatomy and Physiology." Edited by R. B. Todd, M.D., F.R.S.

General Works on the Microscopic Anatomy of the Body.

Beale, Lionel S., F.R.S.—"On the Structure of the Simple Tissues of the Human Body, with some Observations on their Development and Growth."

Carpenter, W. B., M.D., F.R.S.—"The Microscope and its Revelations."

Gosse, Philip H., F.L.S.—"Evenings at the Microscope."

Kölliker, Professor Albert—" Manual of Human Histology." Translated by Mr. Busk and Professor Huxley. Syd. Soc.

Kölliker, Professor A.—" Microscopic Anatomy."

Lankester, Edwin, F.R.S.—" Half Hours with the Microscope."

Quain, Dr.—" Elements of Anatomy." Edited by Wm. Sharpey, Sec. R.S., and G. V. Ellis, Professor of Anatomy and Physiology in University College, London.

Quekett, Thomas, F.R.S.—" Lectures on Histology."

The Blood.

Harvey, Wm.—The works of, Translated for the Sydenham Society.

Richardson, B. W.—" Essay on the Cause of the Coagulation of the Blood."

Chemistry of Physiology.

Johnston, J. F., M.A.—" Chemistry of Common Life." 2 vols.

Lehmann, Dr.—" Physiological Chemistry." Translated by Dr. Day. Sydenham Society.

Liebig, Baron.—" Animal Chemistry." Translated by Dr. Gregory.

Simon, Dr. F.—" Animal Chemistry." Translated by Dr. Day. Sydenham Society.

Digestion and the Digestive Organs.

Beaumont, Wm., M.D., Surgeon U.S. Army.—" Experiments and Observations on the Gastric Juice and the Physiology of Digestion."

Beale, L. S., F.R.S.—" On some Points in the Anatomy of the Liver of Man."

Gray, Henry, F.R.C.S.—" Structure and Use of the Spleen."

Hassall, A. H., M.D.—" Food and its Adulterations."

Lankester, Edwin, F.R.S.—" Twelve Lectures on Food."

Liebig, Baron—"Chemistry of Food." 1847. "Researches into the Motion of the Juices of the Animal Body."
Owen, Professor, F.R.S.—"Odontography."
Prout, Wm.—"Chemistry, Meteorology, and the Function of Digestion considered, with reference to Natural Theology"—Bridgewater Treatise. 1834.

Ethnology.

Huxley, Professor, F.R.S.—"Man's Place in Nature."
Knox, R., M.D.—"The Races of Men."
Lyell, Sir C., F.R.S.—"The Antiquity of Man."

The Eye.

Bowman, Wm., M.D., F.R.S.—"Lectures on the Parts concerned in the Operations on the Eye."
Brewster, Sir David—"Optics"—Lardner's Cyclopædia.
Lloyd, Dr.—"On Light and Vision."
Mackenzie, Wm., M.D.—"Outlines of Ophthalmology."
Morton, Francis—"Optics."

Forces concerned in Life.

Grove, Mr., Q.C., F.R.S.—"The Correlation of the Physical Forces."
Matteucci, Signor—"Lectures on the Physical Phenomena of Living Beings."

The Nervous System.

Bain, Alexander, M.A.—"The Senses and the Intellect," and "The Emotions and the Will."
Bell, Sir C.—"Researches in the Nervous System," by Alexander Shaw.
Grainger, Mr., F.R.C.S.—"Observations on the Spinal Chord."
Sequard, Dr. Brown, F.R.S.—"Lectures on the Physiology of the Nervous System."
Solly, Mr., F.R.S.—"On the Human Brain."
Todd, R. B., M.D., F.R.S.—"Anatomy of the Brain, Spinal Chord, and Ganglion."

Relation of Physiology to other Sciences.

Whewell, William, D.D.—"History of the Inductive Sciences." "Philosophy of the Inductive Sciences."

On the Skeleton, &c.

Holden, Luther, F.R.C.S.—" Human Osteology."
Humphrey, George, M.D., F.R.S.—" A Treatise on the Human Skeleton."
Maclise, Joseph, F.R.C.S.—" Comparative Osteology; being Morphological Studies, to demonstrate the Archetype Skeleton of Vertebrated Animals."
Owen, Professor, F.R.S., &c., &c.—" On the Archetype and Homologies of the Vertebrate Skeleton."

Skin.

Wilson, Erasmus, F.R.C.S.—" A Practical Treatise on Healthy Skin."

INDEX.

A.

ABDOMEN, 11, 12.
Abernethy, anecdote of, 41.
Aberration, spherical and chromatic, 130.
Absorption, 44; experiments on, 42.
Acid, carbonic, where formed, 80; poisonous effect of, 81; of the stomach, 35.
Adam's apple, 83.
Affinity of air-cells for carbonic acid, 59; tissues for blood, 58.
Ague, phenomena of, 91.
Air, action of plants on, 3; cells of lungs, 71; composition of, 80; consumption of, per annum, 81; mixed with food by saliva, 33; influence of, on blood, 79; quantity of, changed in lungs, 77; contained in lungs, 77.
Albumen of blood, 48; egg, 23; porter, 21; potatoes, 43.
Albuminate of iron and soda, 49.
Alcohol, action of, objections to, 20, 21; production of, from oxygen and hydrogen, 13; use of, 21.
Aloes, smallest particle of, that can be tasted, 146.
Amaurosis, 141.
American physician, views of, on digestion, 45.
Ammonia, action of, on fibrine, 48.
Amœba, digestive organs of, 30.
Animals and plants, distinction of, 2.
Animal division of organs, 10.
Anthropotomist, meaning of the term, 65.
Anvil-bone, 148.
Aorta, 51, 54.
Apes, brain of, 119.
Arborescent form of lacteals, 44.
Arm, bones of, 10; mechanical power of, 106, 107.
Arsenic, a poison, 26.
Arterial and venous blood, colour of, 49.
Arteries, elastic coat of, 50; general, 51; pulmonary, 55.
Asphyxia, or suffocation, 70.
Assafœtida, action of, on nose, 143.
Association, British, experiments of, on circulation, 64.

Asthma, 73.
Atropia, action of, on eye, 135.
Auricles, 53; and ventricles, communication between, 54.

B.

BAIRISCH BEER, 21.
Barber-surgeons of England and Germany, 67.
Bath-sponge, advantage of using, 46.
Bath, Turkish, arguments in favour of, 104; injurious effects of, 103; warm, in France, 90.
Battle of physiologists, 17.
Beer, use of, as food, 21; absorption of, 42.
Bees convert sugar into **fat, 23.**
Bell in an air-pump, 150.
Belly. See Abdomen, **11, 12;** and the members, 29.
Bernard, M., experiments of, on use of pancreas, 40; saliva, 33.
Berthelot, discoveries of, 13.
Bile, experiments on use of, 38; composition of, 39; duct, 38; use of the, ib.; where formed, 37.
Biology, derivation of the term, 27.
Bird, experiments on, with carbonic acid, 80; hydrogen, 81.
Bladder-gall, 38; urinary, **102.**
Blade bones, 9.
Blood, circulation of, 47—69; clot of, 47; coagulation of, ib.; colour of arterial and venous, 49; views of Bernard on, ib.; composition of, 48; corpuscles, use and death of, 50; form of, 47; passage of, into capillaries, 52; quantity of, in man, 48; consumed per annum, 81; serum of, 47; temperature of, 99; views of Liebig on, 49.
Blubber, quantity of, **eaten** by Esquimaux, 23.
Blue-pill, how long remains in the system, 26.
Bones. See Skeleton, 7, 8.
Brain, conclusions of Wagner regarding, 119; diseases of, 117; grooves of, 114, 118; of man, smaller than that of elephant, 119; proper, exact use of,

117; quantity of grey matter in, at different ages, 118; relation of, in man (as regards size) to its nerves, 119; man's to that of apes, *ib.*; vertical height of, 120; weight of, in both sexes, 119.
Bread made in Denmark, 27; and water diet, *ib.*
Breath. See Respiration, 70.
Breathe, the reason why we, 82.
Bronchial tubes, 72.
Bronchitis, 89.
Broth, nutritious property of, 26.
Brünner, glands of, 36, 37.
Buffalo-licks, where found, 24.
Butter, acid of, in gastric juice, 35; produced by rheumatism, 97; retailers, practice of, 23.

C.

CÆSALPINUS, discoveries of, 69.
Calmuc and Kaffre, skulls of, 126.
Canal digestive, structure of, 34.
Canals, semi-circular, of ear, 148.
Canine teeth, 31.
Capillaries, by whom discovered, 69; derivation of the word, 50; where they lie, *ib.*
Carbonic acid, action of on albuminate of iron, 49; ditto on blood, *ib.*; influence of on life, 81.
Carnivora, length of intestine among, 41.
Carpenter, Dr., quotation from, 151.
Cataract, 141.
Cell, what is a? 17; theory, explosion of, 17, 18; white, of the blood, 49.
Centigrade thermometer, 90.
Centre of gravity, 108.
Cerebellum, 114; function of, 121; of pigeon, experiments on, *ib.*
Cerebrum, 114, 115, 118, 119.
Cheese, what it is, how differs from albumen, formation of from milk, 23; of vegetables used by Chinese, 24.
Chest, structure of, 73, 74.
Chewing of food, 32.
Chord, spinal, 115.
Chord, vocal, 84; use of, length of, position of, 85, 86.
Choroid coat, use and structure of, 133.
Christiana beer, 2.
Chyle, 44.
Cilia, use and form of, 72, 73.
Ciliary muscle, 135.
Circulation of blood, discovery of, 68, 69; in capillaries, 58; proof of, 64, 65, 66; rapidity of, 61; who coined the term, 69.
Clay mixed with food, 27.
Cleanliness, benefit of, 46.
Clothing, waterproof, why oppressive, 99.
Coagulation, nature of, 48.

Coccyx, 8.
Cochlea of ear, 148.
Coffee, use of, as food, 25.
Cognac, beneficial effect of, 21.
Cold, are we ever, 91.
Collar bones, 9.
Colour-blindness, 138; cause of, *ib.*
Colours, primary, *ib.*
Combustion, nature of, 92; classification of food in connection with, *ib.*
Concha of ear, 146.
Consumption, or Phthisis, 89.
Consumptive children, peculiarities of, 22.
Coracoids, 9.
Corns, growth of, 96.
Cornea, structure of, 133.
Correlation of forces, illustration of the, 13.
Coughing, 88.
Cranium, 9.
Creator, necessity for admitting existence of, 5.
Cricoid gristle, 84.
Crystallization, resemblance of, to growth, 15.
Cuticle, position of, 95.

D.

DAVY, Sir H., anecdote of, 154.
Death, apparent, case of, 15, 16; somatic and molecular, 15.
Decomposition, a proof of death, 16.
Deglutition, 42.
Denmark, peat-formations of, 127.
Dentine of teeth, 31.
Derma of skin, 95.
Development, meaning of the term, 18; of lower plants, 14; of sinew, 18.
Diaphragm, action of, 76.
Diarrhœa, from using gelatine, 52.
Diatomaceæ as food, 27.
Diffusion, mutual, 77, 78.
Digestion, 29; organs of, 31, 32.
Distinction between plants and animals, 2.
Division of natural objects, 1.
Dogs, experiments on, with sugar, 27.
Drowning, how takes place, 70; remarkable case of, 16.
Drum of ear, 147.
Duchek, experiments of, on alcohol, 21.
Dyspepsia, derivation of the word, 46.

E.

EAR, external and internal, 147, 148.
Eggs, composition of, 23.
Elements and words, analogy of, 2 nature of, *ib.*; number of, *ib.*
Enamel of teeth, 31.
Endosmose, 4.
Epidermis, 95.
Epiglottis, 84.

Esquirol, M., opinion of phrenology, 125.
Estimate of man's architecture, 7.
Ether, evaporation of, 99.
Ethnology of skulls, 125, 127.
Eustachian tube, 149.
Exosmose, 4.
Expiration, 76.
Eye, the, 132; ciliary muscle of, 137; coats of, humours of, 134, 135; diseases of, 141; focusing power of, 137; formation of images by, 136; lids, lashes, and brows of, 140; muscles of, 133; pupil and iris of, 135.

F.

FABRICIUS, discovery of, 69.
Fahrenheit's thermometer, 90.
Faraday, Professor, domain of, 91.
Fat, when digested, 35, 36; use of, as food, 22.
Feathers and teeth, relatives, 32.
Fibrine, action of ammonia on, 48; bearing of, on medical jurisprudence, ib.; network of, ib.; relation to albumen, ib.; use of in blood, ib.
Fire-king, the, 94.
Flesh, structure of, 109; when digested, 35.
Food, classification of, 20; definition of, 19; one form of, experiments with, 27; quantity of required per diem, 46; time required for digestion of, 45; to digest, what we require, 29.
Forces, correlation of, 13.
Force, vital, 12.
Freezing of maggots, the, 94.
Friction, production of heat by, 93.
Furnace-function of lungs, 92.
Furrows on the fingers, 96.

G.

GALEN, discovery and period of, 68.
Gall-bladder, 38.
Ganglion, what is a, 116.
Gastric glands, 35; juice, ib.
Gelatine, use of as food, 25.
General practitioners of England, 67.
Germ, action of, on air and water, 14.
German barber-surgeons, 68; theory of cells, 17.
Girdles, locomotive, 9.
Glands, what are, 30.
Glass pounded, a poison, 26.
Gravel, 102.
Gravity, centre of, 108, 109.
Groux, M., breast-bone of, 63.
Grove, Mr., views of, on force, 13.
Gullet, 30.

H.

HAMBURG WHALER, experiences of, 23.
Hammer-bone of ear, 148.

Harvey, discoveries of, 68; simile of, 62.
Heads, round and long, 125.
Hearing, organs of, 146.
Heart, auricles and ventricles of, 53; expansion of, 56; form and structure of, 52, 53; force of, 57; force-pump, action of, 56; impulse of, 63; rhythm of, 61; sounds of, 62, 63.
Heat, action of on solids and liquids, 98; animal development of, 92; great, power of resisting, 94; influence of nerves upon production of, ib.; latent, meaning of the term, 99; maintenance of, 93; nature of, 91.
Heel, action of in walking, 108, 109.
Heel, bones composing, 10.
Hiccups, 90.
Hot and cold, why we feel, ib.
Humboldt, and people of Oronoco, 27.
Humours, vitreous and aqueous, 135.
Hunger, cause of, 27; production of in different animals, 28.
Huxley, Professor, lectures of, 127.
Hygiene, 81.

I.

IDEAL VERTEBRA, 7.
Ilia, 9.
Incisor teeth, 31.
Indians, American, perception of sound by, 151; Peruvian, nasal powers of, 143.
Ingoldsby legends, lines from, 64.
Insane, power of the, to remain without food, 28.
Inspiration. See Respiration, 70.
Integument, 11, 95, 96.
Intelligence, indications of in brain, 34.
Intestines, ib.; length of in sheep, 41; in man, ib.
Iris, use and structure of, ib.
Iron in the blood, 49.
Ischia, 9.
Isthmi-faucium, 42.

J.

JAUNDICE, 38.
Jaw, movements of, 32.
Jellies, value of as food, 26.
Juice, intestinal, 36; experiments on, 37

K.

KIDNEYS, the, 100; membranous canals of, 102; microscopic anatomy of, 101; secretion of, 102.
Knowledge, what consists in, 12.

L.

LACHRYMAL GLAND, 141.
Lacteal absorption. Views of two schools, 44, 45; trunks and leaflets, 44.

Lacteals in the dog, experiments showing, 45.
Lactic acid, of perspiration, 97; experiments with on dogs, 98.
Larynx, 84.
Lassaigne, experiments of on saliva, 34.
Lateral ventricles of brain, 114.
Laughter, production of, 88, 89.
Lens crystalline, 135.
Lenses, action of on light, 129.
Levers, varieties of, 106.
Lewis, G. H., on gelatine, 25.
Liebig, Baron, views of, on alcohol, 22; on beer, 21; on blood, 49.
Life, nature of, 12.
Light, concealment of by plants, 13.
Lighthouses built on a common plan, 7.
Light, rays of, laws regulating, 129.
Lights. See lungs, 71.
Limbs, how moved, 106.
Lithotomy-stool, 104.
Liver, structure and use of, 37, 38.
Locomotion, 106; in girdles, 9.
Lower plants, development of, 14.
Lung, structure of, 71; use of, 79; lobules of, 71.
Lytton, Sir E. B., "Strange Story" of, 16.

M.

MAGENDIE, experiments of, on dogs, 27.
Malpighi, discoveries and period of, 69.
Mammals, what are, 24.
Man, early races of, 126; pre-historic, age of, 127.
Mara, Madame, voice of, 86.
Mariners, shipwrecked, hunger among, 29.
Marrow, spinal, 116; function of, 122.
Mastication of food, 32.
Matter, indestructibility of, 13.
Meals, propriety of resting after, 46.
Meat, decaying, action of, on sugar, 15.
Mechanism of man, method of examining, 5.
Medulla oblongata, 114; function of, 121.
Melancholy and bile, 38.
Melt, or spleen, 104.
Membrane supporting intestine, 41.
Membranes of the brain, 114.
Mental faculties, 118.
Mercury no poison, 26.
Meridian, ethnologic, 126.
Milk, composition and use of, 24.
Minerals, use of, as food, 24; required as food, ib.
Mitral valves, 55.
Molar teeth, 31.
Motion and Locomotion, 106.
Motion, nerves of, 122.
Mucus of intestine, 35.
Muscle, action and contraction of, 110; structure of, 102.

Musk, perfume of, 144.
Mycetes, larynx of, 86.

N.

NAPOLEON's opinion of phrenology, 125.
Nasmyth's steam-hammer, 86.
Nephrophagous, derivation of the word, 103.
Nerve of hearing, 148; of sight, 133; of smell, 143; sympathetic, 116; of taste, 144.
Nerves, termination of, 122, 123; of the intestines, 41; sensory and motor, distinction of, 123.
Nervous system, 111—125; and life related, 111; matter, structure of, 113.
Network of blood-vessels in frog's foot, 68; in kidney, 101; of fibrine, 48.
Nicotiana tabacum, 28.
Nicotine, ib.
Nightshade, deadly, action of on lung, 73.
Nitrogen, use of, in air, 80—83.
Nostrils, 142.

O.

OIL OF TOBACCO, 28.
Olfactory lobes, 114.
Optic nerves, 133.
Optics of the eye, 129.
Organization, meaning of the term, 14.
Organs, definition of, 5.
Orthognathous skulls, 126.
Oscillation in walking, cause of, 109.
Oven, reason why a man can enter with safety an, 99.
Oxygen, action of, on phosphorus, 80; use of, 79.
Ozone, action and composition of, 80.

P.

PACINIAN CORPUSCLES, 152.
Pancreas, 40.
Pancreatic juice, 40; experiments on, ib.
Papillæ of skin, 96; tongue, 98.
Parts, soft, of man, 10.
Pastrycooks, labours of, 22.
Pepys, Samuel, and barber-surgeons, 67.
Perspiration glands, form and number of, 96, 97; fluid, composition of, 97.
Perspire, how we, 98.
Peyer, glands of, 36.
Pharynx, 41; and nostril communication between, 83.
Phosphorus, action of oxygen and nitrogen on, 80.
Phrenology, errors of and objection to, 123—125.
Physiology, importance of, to physician, 64 derivation of the term, 16.

INDEX. 167

Pigment of the eye, 133.
Pivot of levers, 106.
Plan, general, of skeleton, 104.
Plants, benefit of, to man, 2; temperature of, by day and night, 93.
Platinum, powdered, action of, on oxygen, 14.
Pleura of lung, 71.
Poison, what is a, 26.
Poles, barbers', meaning of, 67; ethnologic, 126.
Porter, Dublin, composition of, 21.
Potatoes, absorption of, 42, 43.
Power of levers, 106.
Prehension, 10.
Principle, vital, 12.
Principleists, school of, 16.
Prism, effect of on rays of light, 131.
Prismatic spectrum, 132.
Prognathous faces, 126.
Protective skeleton, 8.
Pubes, 9.
Pulse, beats of, number in different ages, 61; cause of, 59; influence of mind over, 60.
Pupil of the eye, 135.
Pyramidal gristles of larynx, 85.
Pyropathy, meaning of the term, 103.

Q.

QUININE, smallest particle of, that can be tasted, 146.

R.

RACES of man, 126, 127.
Raw material, weaving of, as food, 19.
Reflex actions, nature of, 122.
Refraction, ib.
Reptiles, effect of hunger on, 28.
Respiration, function of, 70—80; influence of external impressions on, 83.
Rest, why bodies remain at, 108.
Retina, structure and use of, 134; paralysis of, 139.
Rheumatism, cause and prevention of, 97, 98.
Ribs, movement of, 75, 76; of belly, 8; of chest, ib.; of neck, ib.
Rickets, from what it proceeds, 24.
Roots of nerves, 122.
Russian Tartary, line drawn from to the Bight of Benin, 126.
Russians, predilection of, for grease, 23.

S.

SALIVA, action of on starch, 34; experiments on, 33; influence of mind over secretion of, 34; quantity of, secreted in 24 hours, 45; quantity of, required for various kinds of food, 34; use of, 33.
Salivary glands, 32, 35.
Salt, importance of as food, 24.

Salts, decomposition of, produces heat, 93.
Sanguis, journey of, 63—65.
Schiller on the four elements, 22.
Sciences, other, relation of to physiology.
Sclerotic coat, structure of, 133.
Scrofula prevented by sugar-eating, 22.
Secretions, glandular, quantity of in 24 hours, 45.
Seeds, influence of heat and light on, 13.
Semilunar valves, 55.
Sensation, nerves of, 123.
Senses, the, 128—155.
Servetus, discoveries of, 69.
Sighing, cause of, 88.
Sign of German phlebotomists, 67.
Sight, long and short, 138; organ of, 132.
Sinew, a, development of, 18.
Skeleton, general character of, 7—10.
Skin, structure and thickness of, 95.
Skull, vertebræ of, 9.
Skulls, shape of, among various nations, 125, 126.
Smell, sense of, 142, 143; nerve of, ib.
Smith's, Sidney, recipe, 91.
Sneezing, how performed, 88; sobbing, ib.
Soda and iron, albuminate of, 49.
Sound, nature and production of, 149, 150; pitch of, 151; reception of by the ear, 149; vibration of, 151.
Sounds of heart, imitation of, 63.
Spectrum, solar, 131.
Speculations, impropriety of forming, 12.
Speech, organs of, 87.
Spinal chord, division of, 115; grey matter of, 116.
Spirits of salts in gastric juice, 35.
Spirometer, 78.
Spleen, function of the, 104, 105.
Sponge-bath, advantage of using the, 46; related to man, 2.
Spot, vital, where situate, 121.
Stammering, cause of, 87.
Stark, Dr., death of, 27.
Stereoscope, explanation of, 139.
Stethoscope, 150.
Stirrup-bone, 148.
Stomach, 34.
Strychnine the food of a bird, 27.
Sugar, passage of, by endosmose, 4; formation of by liver, 38.
Suction, experiments on, 75.
Sulphur of bile, absorption of, 40.
Sweat-glands, character of, 96; number and action of, 97.
Sweetbread, characters of, 40.
Sympathetic system, 116.

T.

TASTE, organs of, 144; varieties of 146.
Tea, use of, as food, 25.
Tear-gland, 141.

Tears, **secretion** and use of, 141.
Teeth, **number,** structure, and **variety** of, 31.
Teetotalism, absurdities of, 20.
Theatre, German, accident in, 68.
Thorn-apple, action of on lung, 73.
Thorax, 73, 74.
Thymus gland, 105.
Thyroid gland, 105; gristle, 83.
Tissue, what is a, 17.
Tobacco, effects of, on the system, 28.
Tongue, 144.
Torricelli's explanation of suction, 75.
Touch, organs of, and experiments on, 151, 155.
Townsend Col., influence of, over his heart, 64.
Trachea, or windpipe, 71.
Tree-lacteal, the, 44.
Tricuspid valves, 55.
Turkish Bath, action of, 103.
Tympanum, 147.
Typical vertebra, 7.

U.

UREA, where, is formed, 102; conversion of gelatine into, 26.
Urine, composition of, 102; secretion of, ib.

V.

VACUUM, nature of, 75.
Valves, neutral, 55; semilunar, ib.; tricuspid, ib.; of veins, 50, 58, 59.

Varolii Pons, 115.
Vegetables, **action of,** on air, 3, 14.
Vegetal division of organs, 10.
Ventilation, **importance of attention to,** 81, 82.
Ventricle of heart, 54.
Ventricles, lateral, of brain, 114.
Ventriloquism, views **as to the nature of,** 87.
Vertebræ, 7.
Vertebral column, 8.
Vesalius, discovery and misfortunes of, 68.
Vestibule of ear, 148.
Villi, what they are, 35, 44.
Vinegar, in gastric juice, 35.
Voice, **organs of,** 83—87; compass of, 86.

W.

WAGNER, researches of, upon the brain, 119.
Walking (see Locomotion), 108, 109.
Water, action of heat on, 98, 99.
Warm-blooded, 90.
Watch, comparison of, with man, 5.
Weight, our conceptions of, 154.
Wheel-animalcules, boiling of, 94.
Whistling, 87.
Wind-pipe, 70, 73.

Y.

YAWNING, 88.

www.ingramcontent.com/pod-product-compliance
Lightning Source LLC
Chambersburg PA
CBHW031448160426
43195CB00010BB/898